.

SOIL SALINITY AND WATER QUALITY

Soil Salinity and Water Quality

Ranbir Chhabra

A.A. BALKEMA/ROTTERDAM/BROOKFIELD/1996

ISBN 90 5410 727 8

Distributed in USA and Canada by: A.A. Balkema Publishers, Old Post Road, Brookfield, VT 05036, USA.

Foreword

The world-over, concerted efforts are being made to enhance agricultural production to feed and improve upon the living standards of the burgeoning population. Many strategies are in action to enhance the rate of growth in the agricultural sector. One important component is to enhance irrigated areas coupled with efficient water-use in order to achieve sizeable leap in the production of foodgrains, fibre and forest plants. However, the injudicious use of this otherwise expensive input is often associated with development of waterlogging, soil salinity, sodicity and many other soil and environmental degradation processes. Besides giving diminishing returns over time, the water-use efficiencies in many of the irrigation projects had been far below the expectations.

This book presents in detail various principles and practices, translated in adaptable technologies, the adoption of which can lead to preventive measures to control waterlogging, soil salinity and package of practices to reclaim wastelands degraded to various levels. It also deals with economically viable and scientifically sound methods of exploiting marginal underground waters to increase production in those areas where good quality water is scanty. In addition, the book also deals with the emerging issues of environmental degradation, especially the methodologies to put to productive uses urban and industrial wastewaters.

Dr. Ranbir Chhabra, who for the last over two decades has been an active researcher, teacher and extension worker in this vital area of reclamation and management of salt-affected soils, has very ably synthesised the research and field experiences in this book. The interaction with the postgraduate students of the Center for Irrigation Engineering of the Katholieke Universiteit Leuven, Leuven, Belgium, where Dr. Chhabra taught a course on Soil Salinity and Water Quality for the last seven years, was a great source of stimulation in writing this book.

I am sure, the vast practical information provided will be of immense use to the students of Soil and Crop Management, Agronomy, Irrigation Engineering, Water Management and to all others directly or indirectly concerned in understanding the principles and will facilitate operationalisation of various options to take not only preventive measures but also to maximise production from this huge but presently degraded resource.

Professor Jan Feyen
Director

Preface

There has been a phenomenal increase in grain production and other agricultural products to support the ever increasing population of the world. Development of irrigation, new varieties, chemical fertilizers and allied agricultural technology has helped many countries to change from deficient to surplus nations. However, this development has not been an unmixed blessing. In spite of best efforts put in, adverse effects of irrigation in the form of waterlogging, soil salinity, alkalinity and overall environmental degradation have been observed in many irrigation commands. In some cases these can be mitigated by efficient water management and introduction of preventive measures. In others, remedial measures in the form of reclamation would be necessary to bring back sick soils and to restore their productivity for sustainability of irrigated agriculture.

With recent advances in knowledge and the development of new concepts, a strong need was being felt for a comprehensive book giving details about the extent, nature and properties of the soils and scientific and practical information about the various technologies available for preventing the formation and reclamation of such wastelands.

This book deals in detail with various factors leading to the genesis of salt-affected soils and the geographical distribution through different continents. Various systems of classification as followed in literature have been dealt with in detail with greater emphasis and justification of the system recently developed in the Indian subcontinent. Emphasis has been given to anthropogenic factors in the development of soil salinity particularly in the developing countries where human pressure is more obvious due to various reasons. Scientific management of salt-affected soils plays a major role in reclamation of these soils and increasing biological productivity of these habitats. Due emphasis has been given for maintaining salt balance, removing of salts by leaching through flooding, drainage or adopting judicious means of irrigation and applying suitable amendments.

In recent years global attention is being given to develop non-conventional crops for salt-affected and waterlogged situations. These aspects have been reviewed with utmost care and vast data related to salt tolerant crops is provided. The mechanism of salt tolerance has been elucidated for the better understanding of students of Plant Physiology and Crop Sciences. Very little, and that too, scattered information in available in the literature about genesis, formation, reclamation and management of alkali soils, particularly for raising crops and trees. Choice of suitable crops and crop sequence, along with judicious fertilizer management based on alkalinity, salinity, nutrient interactions, antagonism among cations and anions, has been explained together with the role of organic manures in increasing and sustaining the productivity of reclaimed salt-affected soils.

Saline agriculture using brackish water is gaining global importance especially in arid areas and along coasts. Criteria for judging the suitability of groundwaters to exploit those for augmenting irrigation potential of the area have been dealt with in detail together with the possibilities of using sewage and industrial effluents for increasing biological production.

This book will be useful to students of Soil Science, Agronomy, Plant Physiology, Environmental Science, Scientists, Planners and Technologists in developed and developing countries.

I take this opportunity to thank Prof. Jan Feyan, Director, Centre for Irrigation Engineering, Catholic University of Leuven, Belgium for his whole hearted support and encouragement for writing this book. Constructive suggestions of Prof. A. Cremer, Prof. R. Dudal and Dr. D. Lamberts of the Faculty of Agriculture, KUL, Belgium are thankfully acknowledged. Contributions of Mrs. Greta Camps, Secretary, and Mrs. Nina at the Centre for Irrigation Engineering for typing the manuscript are duly acknowledged. Assistance of Dr. N.P. Thakur, in editing the manuscript deserves special mention. The constant encouragement of my wife Dr. Aruna Chhabra and the patience of my children Dhiraj and Anuj in allowing me to spare time for this work is duly acknowledged.

Ranbir Chhabra

List of Abbreviations

BH	breast height, 1.35 m from the surface
BOD	biological oxygen demand, mg/l
CEC	cation exchange capacity, me/100 g soil
dS/m	deciSiemens per metre
DW	drainage water
EC	electrical conductivity
ECe	EC of saturated soil paste extract, dS/m
EC_{DW}	EC of drainage water
EC_{IW}	EC of irrigation water
ES	exchangeable sodium, me/100 g soil
ESP	exchangeable sodium percentage
ESR	exchangeable sodium ratio
ET	evapotranspiration
FC	field capacity
FYM	farmyard manure
GM	green manure
GR	gypsum requirement, t/ha
ha	hectare
IW	irrigation water
kg	kilogram
l	litre
LF	leaching fraction
LR	leaching requirement
LSD	least significant difference
me/l	milliequivalent per litre
mg	milligram
OM	organic matter
OP	osmotic pressure, bars
pHs	pH of saturated soil paste

q	quintals
RSC	residual sodium carbonate, me/l
SAR	sodium adsorption ratio
SP	saturation percentage
t	tonne
TDS	total dissolved solids
WP	wilting point

Contents

1

Origin and Distribution of Salt-affected Soils

1.1. General Aspects

All soils contain some amount of soluble salts. Many of these salts act as a source of essential nutrients for the healthy growth of plants. However, when the quantity of the salts in the soil exceeds a particular value, the growth, yield, and/or quality of most crops is adversely affected, to a degree depending upon the kind and amount of salts present, the stage of growth, type of plant, and environmental factors. Thus, soil that contains excess salts so as to impair its productivity is called salt-affected soil.

Land is finite and the ever-increasing demand for its products has put this ecologically vulnerable resource under great stress. Whereas compulsions to expand crop area have brought arable farming to lands otherwise unsuitable for crop cultivation, intensive agriculture supported by irrigation has rendered many productive soils infertile. Continuous depletion of nutrients from soils, waterlogging, and secondary salinisation are some of the attendant problems threatening sustainability of crops in irrigated areas. Owing to these degradation processes, large areas of otherwise productive lands have already gone out of production or are producing sub-optimal yields. In many areas the problem is latent and could assume serious proportions if proper care is not taken to control the rise of water table upon the introduction of irrigation.

1.2. Impact of Salt-affected Soils

Salts not only decrease the agricultural production of most crops, but also, as a result of their effect on soil physicochemical properties, adversely affect the associated ecological balance of the area. Some of the harmful impacts of salts are:
   a)   low agricultural production;

b) soil erosion, by both water and wind, due to high dispersibility of soil and decrease in shear stress;

c) increase in floods due to higher runoff as a result of decreased permeability of soil;

d) low groundwater recharge;

e) change in marine life forms from fresh water to brackish water;

f) ecological imbalance due to change in plant cover from mesophytes to halophytes, from trees to bushes, etc.;

g) poor human health due to (1) toxic effect of elements such as F, B, and Se, and (2) frequent outbreak of malaria and other diseases;

h) low economic returns due to high cost of cultivation, decreased yields, and poor quality;

i) higher maintenance cost and short life of buildings, roads, dams, tubewells, and farm machinery, which get corroded by high salts and the specific effect of sodium and certain other elements.

1.3. Area of Saline and Alkali Soils

Exact information on area and degree of deterioration is not available for all countries, but a number of estimates have been made on a global basis. Dregne (1977) estimated that 2 and 2.1 billion ha were affected by salinisation and waterlogging, respectively. Massoud (1974) made an estimate of 932 million ha, of which 316 million ha are in developing countries. Balba (1980b) estimated from the desertification map of the world (FAO, 1977) that total area subjected to salinisation and sodification was about 600 million ha. Of these, 45 and 8 million ha of land can be brought under cultivation by improving irrigation systems and drainage, respectively. According to Dudal and Purnell (1986), salt-affected soils occupy nearly 7 per cent of the world's land area. The extent of salt-affected soils in different parts of the world is presented in Table 1.1.

Table 1.1.   Estimate of salt-affected soils in the world (area in 1000 ha)

North America	15,755
Mexico and Central America	1,965
South America	129,163
Africa	80,436
South Asia	85,110
North and Central Asia	211,448
Southeast Asia	19,983
Australia	357,568
Europe	50,749
Total	932,185

Source: Massoud (1974).

1.4. Geographical Distribution

Salt-affected soils are found under varied conditions of soil, climate, and physiography and are important constraints in agricultural production in many countries and in every continent (Fig. 1.1–1.11, Klein, 1986). Their distribution and estimated area is given in Table 1.2 and discussed below.

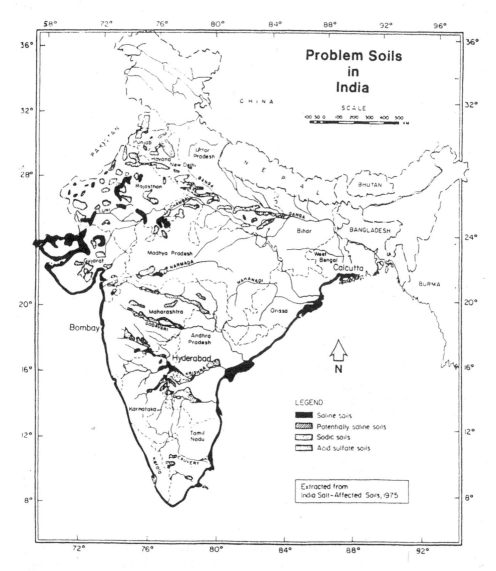

Fig 1.1   Problem soils in India (Extracted from: India Salt-affected Soils, CSSRI Karnal, 1975)

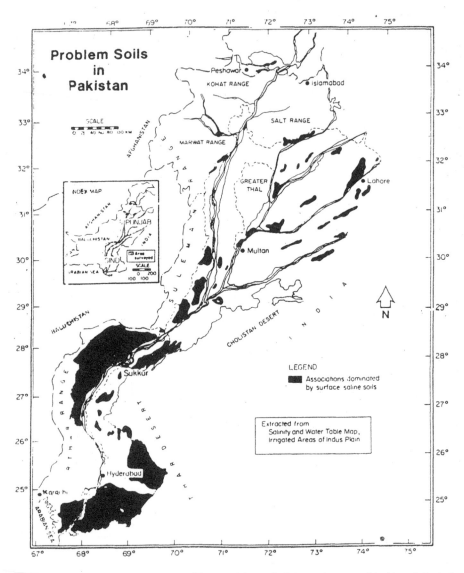

Fig. 1.2. Problem soils in Pakistan (Extracted from: Salinity and water table Map, Irrigated Areas of Indus Plain)

1.4.1. Africa

It has been estimated (Fournier, 1965) that about 2 per cent of the land area of Africa is salt-affected. In the North African region, Margat (1961) stated that salt accumulation is the main cause of low agricultural production in

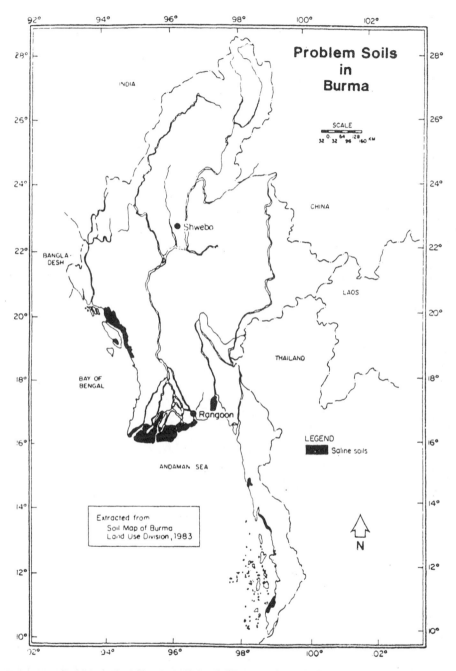

Fig. 1.3.   Problem soils in Mayamar (Burma) (Extracted from: Soil Map of Burma. 1983. Land Use Division, Ministry of Agriculture and Forests, Rangoon, Mayamar).

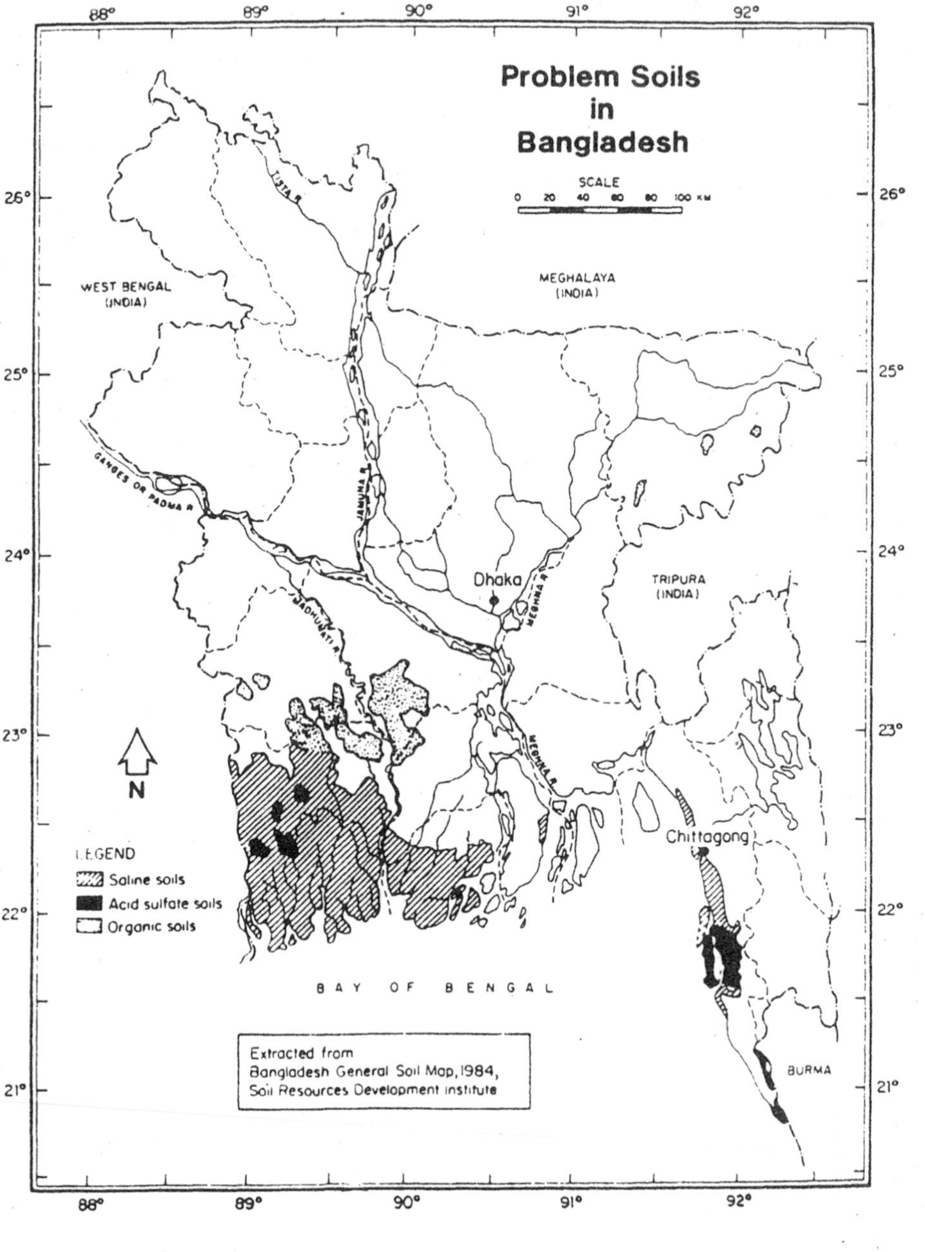

Fig. 1.4.   Problem soils in Bangladesh (Extracted from: Bangladesh General Soil Map, 1984, Soil Resource Development Institute, Dhaka).

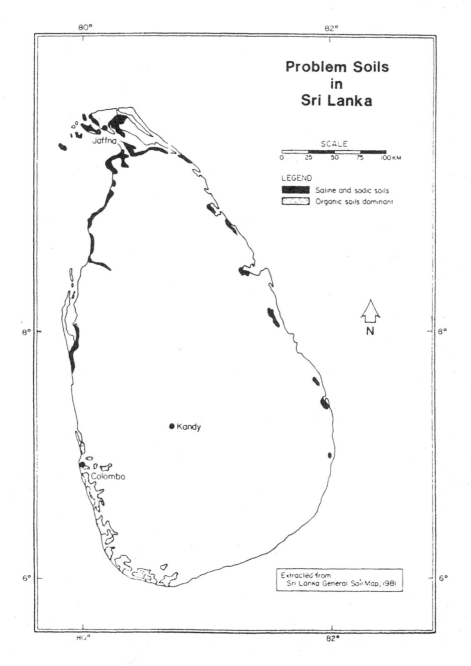

Fig. 1.5.   Problem soils in Sri Lanka (Extracted from: Sri Lanka General Soil Map, 1981, Klein, 1986).

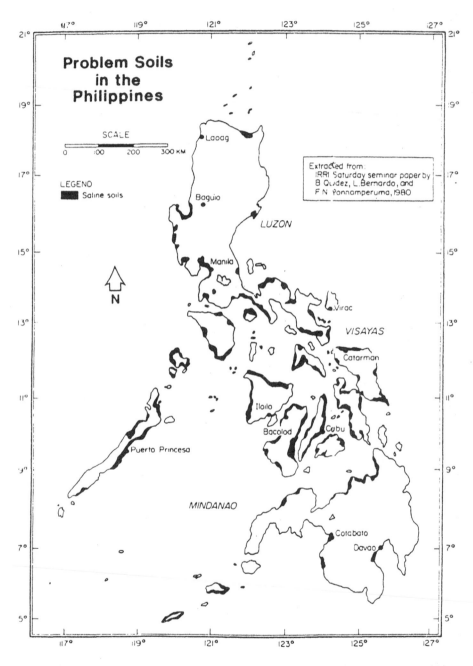

Fig. 1.6.   Problem soils in the Philippines (Quidez et al., 1980).

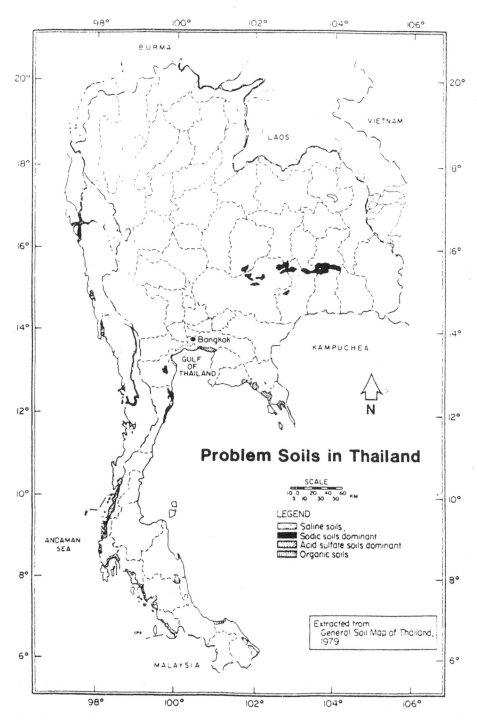

Fig. 1.7.   Problem soils of Thailand (Klein, 1986).

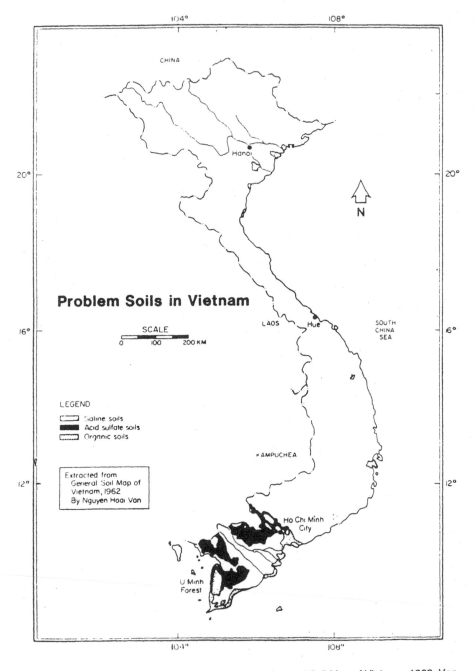

Fig. 1.8.   Problem soils in Vietnam (Extracted from: General Soil Map of Vietnam, 1962, Van Hoai, 1973).

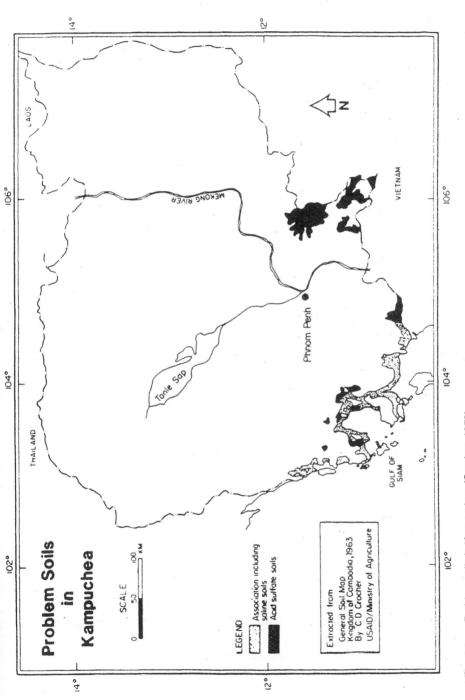

Fig. 1.9.   Problem soils in Kampuchea (Crocker, 1973).

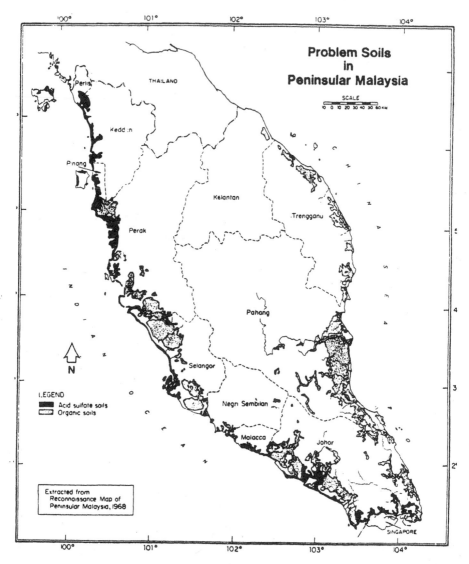

Fig. 1.10.   Problem soils in Peninsular Malaysia (Extracted from: Generalized Soil Map of West Malaysia, 1970, Ministry of Agriculture and Fisheries, Kuala Lumpur).

Morocco. Aubert (1962) cited by Fournier (1965) stated that many of the Sierozems of Morocco, Algeria, and Tunisia have high salt contents and contain layers of gypsum.

In Egypt, primary salinity is widely distributed in the northern lakes region and the Wady-El-Natroun area. Secondary salinisation due to inefficient

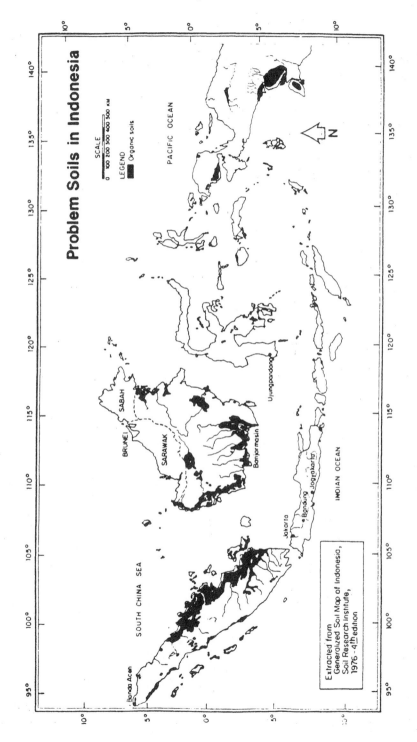

Fig. 1.11.   Problem soils in Indonesia (Extracted from: Generalised Soil Map of Indonesia, Soil Resource Institute, 1976).

Table 1.2.   World distribution of salt-affected soils

Continent	Country	Area, 1000 ha		Total
		Saline/ Solonchaks	Alkali/ Solonetz	
North America	Canada	264	6,974	7,238
	USA	5,927	2,590	8,517
Mexico and	Cuba	316		316
Central America	Mexico	1,649		1,649
South America	Argentina	32,473	53,139	85,612
	Bolivia	5,233	716	5,949
	Brazil	4,141	362	4,503
	Chile	5,000	3,642	8,642
	Colombia	907		907
	Ecuador	387		387
	Paraguay	20,008	1,894	21,902
	Peru	21		21
	Venezuela	1,240		1,240
Africa	Afars and Issas	1,741		1,741
	Algeria	3,021	129	3,150
	Angola	440	86	526
	Botswana	5,009	670	5,679
	Chad	2,417	5,850	8,267
	Egypt	7,360		7,360
	Ethiopia	10,608	425	11,033
	Gambia	150		150
	Ghana	200	118	318
	Guinea	525		525
	Guinea-Bissau	194		194
	Kenya	4,410	448	4,858
	Liberia	362	44	406
	Jamahiriya	2,457		2,457
	Madagascar	37	1,287	1,324
	Mali	2,770		2,770
	Mauritania	640		640
	Morocco	1,148		1,148
	Namibia	562	1,751	2,313
	Niger		1,389	1,389
	Nigeria	665	5,837	6,502
	Rhodesia		26	26
	Senegal	765		765
	Sierra Leone	307		307
	Somalia	1,569	4,033	5,602
	Sudan	2,138	2,736	4,874
	Tunisia	990		990
	United Rep. of Cameroon		671	671
	United Rep. of Tanzania	2,954	583	3,537

(*Contd.*)

Table 1.2. (*Contd.*)

Continent	Country	Area, 1000 ha		Total
		Saline/ Solonchaks	Alkali/ Solonetz	
	Zaire	53		53
	Zambia		863	863
South Asia	Afghanistan	3,103		3,103
	Bangladesh	2,479	538	3,017
	Burma	634		634
	India	23,222	574	23,796
	Iran	26,399	686	27,085
	Iraq	6,726		6,726
	Israel	28		28
	Jordan	180		180
	Kuwait	209		209
	Muscat and Oman	290		290
	Pakistan	10,456		10,456
	Qatar	225		225
	Sarawak	1,538		1,538
	Saudi Arabia	6,002		6,002
	Sri Lanka	200		200
	Syrian Arab. Rep.	532		532
	United Arab Emirates	1,089		1,089
North and Central Asia	China	36,221	437	36,658
	Mongolia	4,070		4,070
	Former USSR	51,092	119,628	170,720
South-east Asia	Democratic Kampuchea	1,291		1,291
	Indonesia	13,213		13,213
	Malaysia	3,040		3,040
	Socialist Rep. of Vietnam	983		983
	Thailand	1,456		1,456
Australia	Australia	17,269	339,971	357,240
	Fiji	90		90
	Solomon Islands	238		238
Europe	Czechoslovakia	6	15	21
	France	175	75	250
	Hungary	2	385	387
	Italy	50		50
	Rumania	40	210	250
	Former USSR	7,546	21,998	29,544
	Yugoslavia	20	235	255

Source: Szabolcs (1974), for data on Europe; Massoud (1974) for all other data.

drainage affects very large areas in and adjacent to the Nile Valley. According to El-Elgabaly (1959), soil salinity and sodicity are major problems in Egyptian agriculture, affecting about 800,000 ha of cultivated soils. In Sudan, soil salinity is a major problem in the Gezira and in Northern Province (Ayoub, 1960).

1.4.2. Asia

Among the countries of South and South-east Asia, the problem is greater in India, Indonesia, Malaysia, Bangladesh, Kampuchea, Thailand and Vietnam. In India, estimates of the total area affected range from 6.1 million ha (Raychaudhuri, 1965) and 7 million ha (Abrol and Bhumbla, 1971) to 23.8 million ha (Massoud, 1974). Only a small proportion is of coastal origin (Raychaudhuri, 1965), while 40 per cent lies in the Indo-Gangetic plain, which is the most productive area of the country. Recently singh (1992) has put this figure to 8.5 million ha. Nazir (1965) stated that in Pakistan there are about 2 million hectares of salt-affected soils in the Indus Valley and about 1.5 millions in the Punjab.

In China, according to Syun, quoted by Bernstein and Hayward (1962), there are 20 million ha of salt-affected soils. Massoud (1974) puts this at 36.7 million ha. Kovda (1965) stated that salinity is widespread in the Sungari Valley in Manchuria and in the Hwang-Ho Delta, particularly near the Yellow Sea. These also occur in many arid and semi-arid regions of the country. Salinity originating from marine chlorides is common near the coast.

Saline and alkali soils occur widely in the Asiatic USSR, with large areas in the river valleys of eastern and western Siberia in the Urals region and in the Araxes Valley in Armenia (Kovda, 1965).

The Middle Eastern countries are much affected by salinity. In Syria it is found in the Palmyra region and in the Euphrates, El-khabour and Dan valleys (Muir, 1951). Salinity has been a problem in Iraq since ancient times, affecting most of the Euphrates and Tigris valleys, whether irrigated or not (Buringh and Edelman, 1955; Dieleman, 1963; Ressel *et al.*, 1965; Sehgal and Sys, 1980). The areas of salt-affected soils in Afghanistan are saline rather than alkali. The largest areas of salt-affected soils in this region, however, are in Iran and Saudi Arabia.

1.4.3. Australia

Salt-affected soils are widespread in many parts of Australia. Both as naturally occuring sodic, and saline soils and those formed as a result of man's activities in sensitive dry land situations. These soils are not widely cultivated and left as natural reserves. Inland salinity is an increasing problem in irrigated areas. Massoud (1974) estimated that 17,269 and 339,971 hectares of saline and sodic soils, respectively occur in Australia. Later on Stoneman (1978) has put the area under saline soils to 167,000 hectares.

1.4.4. Europe

Salt-affected soils are found in the alluvial regions of the Danube, Dnieper, and Don in the former USSR. They also occur in eastern Europe in Hungary, Czechoslovakia, Rumania, and Yugoslavia, where they account for 10 per cent of land area (Szabolcs, 1965). In western Europe, Spain has about 600,000 ha in Andalusia, Aragon, and Catalonia (Ayers and Wadleigh, 1960). The Netherlands are a special case in that though much of the land was originally under sea water and thus saline, it has subsequently been drained and leached free of salts through a system of polders and intensive drainage.

1.4.5. North America

Salt-affected soils are found in 17 western states of the United States of America, notably in Arizona, California, New Mexico, Texas, and Utah, and in parts of western Canada. There are significant areas of saline soils of varying origin in Cuba (Blazhnil, 1957, cited by Bernstein, 1962) and in Mexico.

1.4.6. South America

Salt-affected soils are found in most countries of South America, particularly in Argentina, Brazil, and Chile. Of some interest, though affecting a relatively small area, is the salinity in the coastal belt of Peru, 2000 km long and 10 to 25 km wide, initially of marine origin but now becoming steadily more extensive because of waterlogging due to irrigation and poor drainage facilities. Zavaleta (1965) estimated that crop production was reduced by salt accumulation in 25 to 30 per cent of this belt.

1.5. Nature of Salts

The chief ionic combinations that give rise to the occurrence of salts originate with in the series Ca^{2+}, Mg^{2+}, Na^+, K^+, Cl^-, SO_4^{2-}, HCO_3^- and CO_3^{2-}. These elements belong to the top 15 elements found in the earth's crust, as illustrated in Table 1.3.

Table 1.3. Common element content (per cent) of the earth's crust

Element	% content	Element	% content
Oxygen (O)	49.13	Hydrogen (H)	1.00
Silicon (Si)	26.00	Titanium (Ti)	0.61
Aluminium (Al)	7.45	Carbon (C)	0.35
Iron (Fe)	4.20	Chlorine (Cl)	0.20
Calcium (Ca)	3.25	Phosphorus (P)	0.12
Sodium (Na)	2.40	Sulphur (S)	0.10
Magnesium (Mg)	2.35	Manganese (Mn)	0.10
Potassium (K)	2.35		

Their presence in solutions circulating the earth, in the ocean, and in the continental and in marine deposits can therefore be ascribed, at least in part, to the weathering of crystalline rocks and minerals. These elements can formally be divided into several categories according to their geochemical mobility subsequent to weathering processes (hydrolysis, hydration, oxidation, and carbonation). The mobility sequence in increasing order is as follow:

a) Practically non-leachable   Si in quartz
b) Slightly leachable   Fe, Al, Si, P
c) Leachable   Si, P, Mn
d) Highly leachable   Ca, Na, K, Mg
e) Very highly leachable   Cl, B, I, S, C

Salt formation is therefore expected to result from a combination of the groups d and e to a variety of possible compounds such as $NaCl$, Na_2SO_4, $MgCl_2$, $MgSO_4$, $CaCl_2$, $CaSO_4$, Na_2CO_3, $NaHCO_3$, $MgCO_3$, and $CaCO_3$. Their differential geochemical mobility within a close area may explain the cause of slick spots formation in otherwise normal soils.

1.6. Origin of Salts

Soil salinisation may originate from a variety (and combination) of frequently interrelated sources. However, weathering of rocks and minerals in the earth's crust is the chief (primary) source of all soluble salts present in the soil and sea. Although the salts currently occurring in the ocean arise mainly from the weathering processes of the earth crust, the ocean now functions as an important "source term" for redistribution of salts. The main origin of salts for a particular area can be any one or combination of the following:

a) Soil weathering process
As a result of weathering process, salts are formed in the soil. But under humid conditions salts percolate (leach) through the soil and are transported to the sea by streams and rivers. Therefore, inland salt-affected soils are rarely formed in humid areas. But under arid and semi-arid conditions these weathering products accumulate *in situ* and result in the development of salinity or alkalinity. This process of formation of salt-affected soils as a result of accumulation of salts released during wealtheing is called primary salinisation.

b) Accumulation on the surface due to irrigation under inadequate drainage
As a result of irrigation, the water transports the salts present in the soil profile to the surface and leaves them behind on evaporation. Thus, over a period of time, salts that were previously evenly distributed in the whole profile may selectively accumulate on the surface and give rise to saline soils.

Accumulation of salt-laden runoff water and its subsequent evaporation in the undrained basins is the cause of salinity in many low-lying areas.

c) Irrigation with salt-laden underground water
A fairly common occurrence in arid and semi-arid conditions is the presence of salt-laden underground waters that are increasingly being exploited for irrigation purposes. These are a direct source of salts on otherwise good quality lands. A striking example is the use of high sodium waters, which may lead to poor permeability and sodication of soil.

d) Shallow water table
Under inadequate drainage and inappropriate management of water, during both transport from dams and canals and on-farm use, the water table rises following introduction of irrigation into an area. In several of the irrigation command areas, the water table rises even at a rate of 1–2 m per annum (Fig. 1.12) has been reported. Such groundwaters are often mineralised to some extent, and as a result of capillary effects, water continuously rises upward

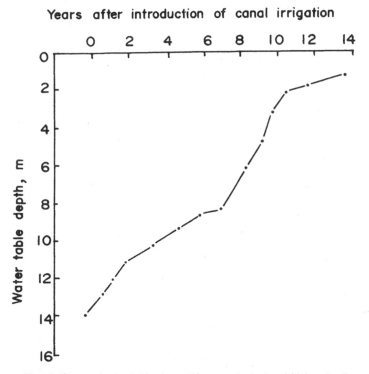

Fig. 1.12. Trend of groundwater table rise at Haryana Agricultural University Farm at Hisar, India, after the introduction of canal irrigation.

and enriches the surface soil in salts following evaporation. This is perhaps the major cause of development of salty lands in irrigated areas.

The salinisation risk of high water table (which may also rise from other causes) is, however, related to a combination of salinity and depth. It is generally only after some "critical depth" that these effects occur. Knowledge of this phenomenon is of course quite important for irrigation and drainage operations in arid zones and is discussed in detail in Chapter 2.

e) Fossil salts

Salt accumulation in arid regions often involves "fossil salts", derived from earlier deposits or entrapped solutions in former marine deposits. Salt release may occur naturally or result from human activities. An example of the former is the rise of salt bearing groundwater through an originally impervious cap (which becomes permeable as a result of weathering processes) overlying saline strata. Examples of the latter are the building of canals or water works in saline strata and the use of groundwater for irrigation.

In Rajasthan, India, a canal built on an underlying gypsum layer has resulted in development of salinity in the area within only a few years of its construction. This has been due to the perched water table and contribution of salts from the underground gypsum layer.

f) Seepage from the upslopes containing salts

Under certain situations, seepage resulting from water inflow in the upslope areas can cause severe salinity of the downslope areas especially when the sub-surface water flow takes place through the strata that are rich in salts and/or marine deposits. Mineral springs owe their salts to such a situation.

g) Ocean

In the coastal areas, the soils generally get enriched with salts from the sea through:

    i)   inundation of surface soil by sea water during high tide;

    ii)   ingress of sea water through rivers, estuaries, etc.;

    iii) groundwater inflows; and

    iv) salt-laden aerosols, which can be transported even many kilometres inland from the seacoast and deposited as dry "fall-out" or "wash-out" by showers. Inland deposition of NaCl at a rate of 20–100 kg/ha/year are quite common and values of 100–200 kg/ha/year for nearby coastal areas have been reported. Although this amount may appear small, regular deposits over a long period may lead to salinisation of the area.

h) Chemical fertilisers and waste materials

Though the use of chemical fertilisers, which are inorganic salts, and manures in agricultural fields is increasing, yet their contributions to the overall salt build-up in soils is insignificant. However, certain situations, such as dumping of cowdung slurry, sewage sludge, or industrial byproducts like pressmud or pyrites, can contribute to excessive accumulation of certain ions that can limit soil productivity.

1.7. Classification of Salt-affected Soils

Based on pH of the saturated soil paste (pHs), total soluble salts or electrical conductivity of the saturated soil paste extract (ECe), and the exchangeable sodium percentage (ESP), various attempts have been made to classify salt-affected soils into different groups. Limits proposed for various groups or classes are discussed below.

1.7.1. The USDA System

The USDA Salinity Laboratory Staff (Richards, 1954) classified these soils into three distinct categories (Table 1.4): saline, sodic, and saline-sodic soils. Since 50 per cent reduction in yield takes place at ECe of 4 dS/m for most agricultural crops, this was proposed as the critical value to distinguish saline from non-saline soils. Similarly, as the physical properties, especially the permeability of the soil, was significantly affected at ESP more than 15, that was taken as a critical value for differentiating sodic from normal soils.

Table 1.4. USDA Salinity Laboratory Staff system of classification

Type of soil	ECe, dS/m	ESP	pH
Saline	> 4.0	< 15	< 8.5
Sodic	< 4.0	> 15	> 8.5
Saline-sodic	> 4.0	> 15	> 8.5

Source: Richards (1954).

Later on, these types of soils were identified under the taxonomical order Aridisols, an order characteristic of arid regions with a light-coloured surface soil and one or more alluvial horizons. The salorthid group has a saline horizon and corresponds approximately to the solonchaks of the FAO-UNESCO classification. The Arid sub-order with a clay horizon may have a sodium-rich layer (Natragids) or both, i.e., compact clay horizon and sodium-rich layers, and corresponds to solonetz soils. Other orders, particularly Mollisols and Alfisols, may be subject to salinity or sodicity phases where conditions favour these developments.

1.7.2. The USSR System

In former USSR, these soils have been classified into two groups: solonchak and solonetz.

a) *Solonchak soils:* Instead of expressing the salt concentration on the basis of ECe, soil salinity is expressed in the USSR on the basis of salt content as per cent of dry soil weight (Kovda, 1965). On that basis, solonchak soils are defined as those that contain more than 2 per cent soluble salts in the upper 30 cm soil.

Furthermore, depending upon the total salt content and the type of soluble salt, five categories have been distinguished (non-saline, slightly, moderately, and strongly saline, and solonchak). The relationship between total salt content and category depends upon the mix of salts present. Thus, depending on type of salts, slightly saline soils may contain from 0.1–0.2 to 0.3–0.6 per cent total salts, while solonchaks contain salts varying from 0.5 to 2.0 per cent.

b) *Solonetz soils:* Solonetz soils are defined as those characterised by a high ESP in the B horizon. Mainly based on the type of clay minerals and ESP, the solonetz soils have been divided into four categories (Table 1.5).

Table 1.5.   USSR classification of solonetz soils

Category	ESP	
	Chernozem soils	Chestnut and Brown soils
Solonetz	> 30	> 16
Strongly solonetzic	15–30	10–16
Moderately solonetzic	10–15	5–10
Weakly solonetzic	< 10	< 5

1.7.3. The European System

In addition to the limits provided for ECe, ESP, and pHs by the USDA Salinity Laboratory Staff, the Europeans (Szabolcs, 1974) introduced in 1968 a genetic parameter, i.e., structure of B horizon, and further classified the two major groups into :

a) saline soil with or without structural B horizon
b) sodic soils with or without structural B horizon

1.7.4. The Australian System

The Australian scientists have classified the salt-affected soils into saline, sodic, and alkaline soils mainly on the basis of per cent salt content, ESP and pHs of the soils respectively (Table 1.6). Considering that in Australia and probably in many other parts of the world, the hydraulic conductivity problems

Table 1.6.   Australian system of classification

Soil type	Category		
	0	1	2
Saline soils based on per cent salt content	Non-saline < 0.1% NaCl	Saline loam > 0.1% NaCl clay > 0.2% NaCl	Highly saline > 0.3% NaCl in B horizon
Sodic soils, based on ESP	Non-sodic < 6	Moderately sodic 6–14	Highly sodic > 14
Alkaline soils based on pHs	Non-alkaline < 8.0	Alkaline 8.0–9.5	Strongly alkaline > 9.5

start, in vertisols, even at very low ESP (McIntyre, 1979), they use much lower threshold values of ESP to distinguish the problem of sodicity.

1.7.5. The FAO-UNESCO System

In the FAO-UNESCO system (FAO, 1974), the salt-affected soils are grouped into two categories:

a) *Solonchak soils:* This group is characterised by high salinity within 125 cm of the surface. High salinity is defined as an ECe of more than 15 dS/m at some time of the year within 125, 90, or 75 cm of the surface in soils with coarse, medium, or fine-textured top soil respectively; or an ECe of more than 4 dS/m within 25 cm of surface soil. The main sub-units are ascribed as orthic, mollic, takyric, and gleyic.

b) *Solonetz soils:* These soils are defined as having (i) natric B horizon in the upper 40 cm of which ESP is more than 15 or (ii) having more exchangeable $Na^+ + Mg^{2+}$ than Ca^{2+} + exchangeable acidity (at pH 8.2) within the upper 40 cm of the horizon and an ESP above 15 in some subhorizon within 2 m of the surface. Soils without a natric horizon but having an ESP above 6 in some horizon within 1 m of the surface are categorised as alkaline phase. The relationship between the widely used classification systems is given in Table 1.7.

1.7.6. The Indian System

In addition to the parameters proposed by USDA Salinity Laboratory Staff, Indian scientists considered the nature of soluble salts as an important index for grouping of these soils (Bhargava *et al.*, 1976; Bhumbla, 1977). They further argued that a pHs of 8.5 is too high, as the isoelectrical pH for precipitation of $CaCO_3$, at which the sodication process starts, is 8.2 and mostly this pH is associated with ESP of 15 or more (Abrol *et al.*, 1980). Considering these

Table 1.7.   Correlation between the widely used classification systems for salt-affected soils

Basic grouping	FAO-UNESCO classification	USDA classification	USSR classification
Saline	**Solonchak**		
	Orthic solonchak	Salorthid	Fluffy solonchak (non-steppic)
			Crust solonchak
		Salorthidic	Soda solonchak (non-steppic)
	Mollic solonchak	Calciustoll	Fluffy solonchak (steppic)
		Haplustoll	Soda solonchak (steppic)
	Takyric solonchak		Takyrs
	Gleyic solonchak	Halaquept (in part)	Meadow solonchak
Alkali	**Solonetz**		
		Natrargid	
		Nadurardig	
	Orthic solonetz	Natriboralf	Desert-steppe and desert solonetz
		Natrudalf	
		Natrustalf	
		Natrixeralf	
		Natralboll	
	Mollic solonetz	Natriboroll	
		Natrustoll	Steppe solonetz
		Natrixeroll	
		Natraquoll	
	Gleyic solonetz	Natraqualf	Meadow solonetz
	Solodic planosol	Argialboll (in part)	Solod

points they have grouped these soils into two categories (Table 1.8). Based on the nature of the major problem the plant faces for its optimum growth and the strategy to be adopted for their reclamation, the soils previously classified as saline-sodic because of their high pHs, ESP, and ECe are to be treated as either saline or alkali.

1.8. Human Role in the Development of Soil Salinity

Good agricultural lands have become salinised because of several faulty human activities such as:

Table 1.8.   Indian system of classification

Soil characteristics	Saline soils	Alkali soils
pHs	< 8.2	> 8.2
ESP	< 15	> 15
ECe	> 4 dS/m	variable, mostly < 4 dS/m
Nature of soluble salts	Neutral, mostly Cl^- and SO_4^{2-}. HCO_3^- may be present but CO_3^{2-} is absent	Capable of alkaline hydrolysis, preponderance of HCO_3^- and $CO3^{2-}$ of Na^+

a) Construction of roads, dams, canals, and bunds, thereby blocking the natural surface drainage of the area, which leads to waterlogging, rise in water table, and ultimately salinisation.

b) Use of saline underground waters for irrigation without providing adequate drainage.

c) Faulty transport and mismanagement in on-farm use of irrigation water leading to higher seepage, over-irrigation, and ultimately secondary salinisation and alkalinisation. It may be mentioned that "*if the construction of the storage and conveyance system is the end of the beginning in the development of irrigated agriculture, the appearance of soil salinity is the beginning of the end*". Most farmers tend to over-irrigate because of unsure canal water supply, and ignorance about the proper depth and frequency of irrigation leading to rise in water table. Similarly, though some soils may be well drained and non-saline under natural conditions, drainage may not be adequate for irrigation.

d) Change in cropping pattern from forest land to agricultural land (there are reports that flow of spring waters has increased after clearing of forest lands), water sparing crops like millets to high water requiring crops like rice; rainfed agriculture to irrigated agriculture, and increase in intensity of cropping lead to more percolation losses of applied irrigation water, causing rise in water table and thus salinisation of many areas. Leaving land fallow normally encourages more accumulation of salts on the surface.

2

Saline Soils and their Management

2.1. Characteristics of Soil Profile and Saturation Extract

Saline soils, also known as solonchaks, are those that contain appreciable amounts of soluble salts so as to interfere with plant growth. Electrical conductivity of the saturation paste extract of these soils is more than 4 dS/m, pH of the saturation paste less than 8.2, and exchangeable sodium percentage less than 15. Saline soil profile is generally monotonous from the top to the bottom, i.e., parent material. These soils lack structural B horizon and contain very little organic matter (OM), less than 1 per cent. Soluble salts mostly consist of Cl^- and SO_4^{2-} of Na^+, Ca^{2+}, and Mg^{2+}. Bicarbonates may or may not be present. However, carbonates are generally absent (Table 2.1). Depending upon the dominance of anion, these soils may be termed $SO_4^{2-}:Cl^-$ solonchaks, as in Marismas (Spain), or $Cl^-:SO_4^{2-}$ solonchaks, as in the Caspion lowland area (Fig. 2.1).

Table 2.1. Composition of saturated paste extract of saline soils

Soil no.	SP	pHs	ECe, dS/m	Na^+	K^+	Ca^{2+}	Mg^{2+}	Cl^-	SO_4^{2-}	HCO_3^-	CEC, me/100 g soil
						me/l					
1	54.0	7.30	2.2	11	2.2	6.9	7.2	12	13	2.9	16
2	52.0	7.60	26.6	260	5.5	16.1	24.0	110	249	4.2	18
3	76.2	7.20	32.3	252	6.1	36.6	62.8	285	114	3.8	32
4	40.7	7.45	27.9	174	17.3	44.9	72.0	224	104	4.8	16

Total soluble cations >CEC.
Exchangeable Ca > 20.

The chemical analysis of the saturation extract of saline soils (Table 2.1) shows high concentration of soluble Na^+ followed by Ca^{2+} and Mg^{2+}. If the concentration of total soluble cations exceeds that of CEC, and or exchangeable Ca is more than 20 me/100 g soil, then the soils invariably contain gypsum

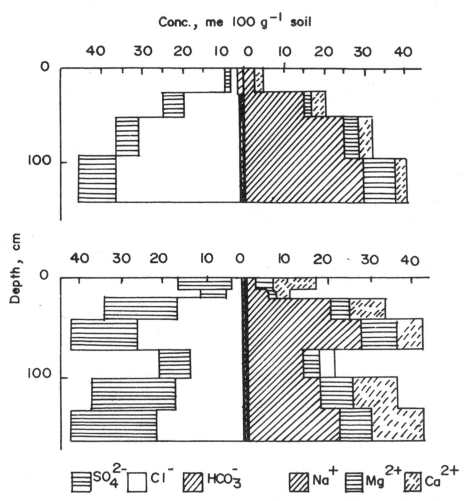

Fig. 2.1.   Salt profiles of saline soils: a sulphate-chloride solonchak of Marismas, Spain (top), and a chloride-sulphate solonchak from the Caspian lowland (bottom).

($CaSO_4$) and are known as gypsiferous saline soils. In addition to high concentration of Cl^-, SO_4^{2-}, and Na^+, saline soils contain toxic concentration of B and Se.

2.2. Development of Saline Soils

Saline soils are formed whenever climate, soil, and hydrological conditions favour accumulation of soluble salts in the rootzone, even temporarily, and are influenced by the following factors:

a) Climate
Except for the coastal areas, saline soils are rarely found in humid regions because there the salts get continuously washed out of the rootzone by the leaching action of rain water. Salt-affected soils are common in arid and semi-arid regions that receive inadequate and irregular precipitation to accomplish leaching of salts originally present in the soil profile. Normally, if the area gets more than 1000 mm rainfall annually, then saline soils are not formed.

The accumulation of the salts in the surface layer can also be enhanced if a cool wet season alternates with a hot dry season. During a dry summer period the upward flux of salts from the groundwater table to the surface can be greater because of their enhanced solubility than the downward flux from the surface layer achieved through leaching during the cool season due to their decreased solubility.

b) Soil
On introduction of irrigation, redistribution of salts, which were previously distributed uniformly throughout the soil profile or were localised in deeper layers, may lead to their excessive build-up in the rootzone. It is well known that, through capillary rise, water brings the dissolved salts to the surface and deposits them there after evaporation. Thus, many soils that are now non-saline carry the risk of becoming saline on introduction of irrigation.

Normally, light-textured soils get less salinised than heavy-textured soils because:

i)   light-textured soils are highly drained and thus the salts get leached out easily and quickly,

ii) these soils have low CEC and thus retain less salts than heavy-textured soils,

iii) light-textured soils have poor capillary rise and thus are likely to be get less affected by shallow brackish water table.

c) Hydrological conditions
The following hydrological conditions may favour waterlogging and/or accumulation of salts resulting in formation of saline soils.

i)   Low-lying areas resulting in surface accumulation of salts brought about by runoff from outside. These areas may look waterlogged during the rainy season but dry out in the off-rainy season, giving rise to external solonchaks. In coastal areas, the inundation of surface soils by sea water during high tides and ingress of sea water through rivers, estuaries, and underground aquifers are major causes of salinity development.

ii) Contributions from the high aquifers and/or adjoining areas through seepage (Fig. 2.2) as described by van der Molen (1976) may increase the total salts and/or raise the water table, encouraging the develop-

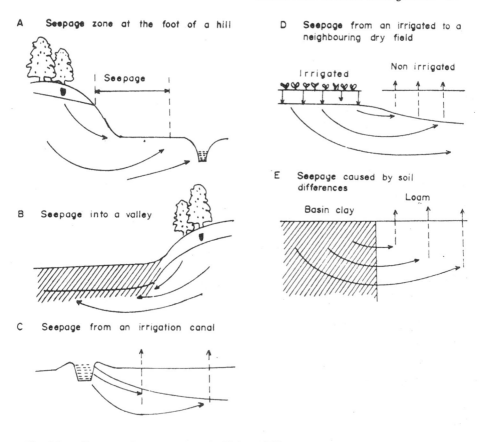

A   Seepage zone at the foot of a hill

Seepage

B   Seepage into a valley

C   Seepage from an irrigation canal

D   Seepage from an irrigated to a neighbouring dry field

Irrigated   Non irrigated

E   Seepage caused by soil differences

Basin clay   Loam

Fig. 2.2.   Seepage phenomena van der Molen, 1976).

ment of salinity. In addition to this, saline seeps, common in Australia, North America, and other countries, may also cause salt accumulation in the lower regions. Doering and Sandoval (1976) and Pluym (1978) reported that a change in land use from a natural forest vegetation to a cereal grain crop or shift in cropping pattern such as the introduction of summer fallow may result in a larger amount of water passing through the soil and carrying salts for the low-lying adjoining areas. Stoneman (1978) stated that salinity problems in western Australia, unlike many throughout the world, are not associated with irrigation but with dryland farming, which reduces the evapotranspiration and thus encourages excess leaching, causing waterlogging and salinity in the lower areas.

iii) Introduction of canal irrigation, which results in:
  — greater total input of water in the area, many times during periods of low water requirement by crops;

— decrease in the groundwater exploitation and thus disturbance in the natural water balance of the area;

— deep percolation losses due to poor on-farm management of irrigation water by the farmers because of their ignorance about the depth and frequency of irrigation and because of heavy irrigation to cover uncertainty about the availability of canal irrigation water;

— bringing unsuitable lands like coarse-textured sandy soils under intensive irrigation.

For these reasons, on introduction of irrigation in an area, the groundwater generally rises or forms a perched water table. Rise in water table is the major cause of soil degradation in many irrigated areas.

d) Mechanism of soil salinisation from shallow water table

The rise of water in the soil from a free water surface, i.e., water table, is termed "capillary rise". This is an important mechanism by which the soil becomes saline because of the upward movement of groundwater and its subsequent transpiration and/or evaporation at the soil surface. While water in the pure state evaporates, salts carried by water are left behind and accumulate in the surface soil (Fig. 2.3). The amount of water and salts that reach the soil surface through capillary rise depend upon the soil texture, depth of water table, i.e., suction head, and concentration of salts in the groundwater.

Mathematically capillary rise is defined as

$$hc = 2\psi \cdot \frac{\cos \theta}{r} \cdot dw \cdot g \qquad (1)$$

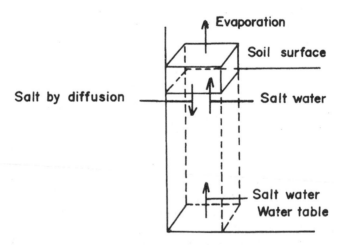

Fig. 2.3.   Flow diagram for the movement of water and salts in a soil having a shallow water table. As water evaporates, salts accumulate at the surface and diffuse downward in response to the concentration gradient that develops (Doering, 1963).

where hc = equilibrium height of capillary rise,

   ψ = surface tension,

   θ = wetting front,

   r = radius of capillary tubes,

   dw = density of water, and

   g = gravitational acceleration.

Keeping all other things constant, this equation predicts that more water will rise in heavy-textured clay soils (finer pores, smaller value of r) than in light-textured sandy soils (wider pores, higher value of r).

Furthermore, steady upward flow of water from water table through the soil to an evaporation zone at the soil surface as described by Darcey's law gives:

$$q = K(\psi) \left(\frac{d\psi}{dz} - 1 \right) \tag{2}$$

where q = flux equal to the evaporation rate E, under steady state conditions,

   $K(\psi)$ = hydraulic conductivity as a function of ψ, and

   z = depth of water table.

This equation predicts that the shallower the water table (z), the higher the upward flux (q) to meet the evaporation demand (E) of the surface soil. Schlesener (1958) observed (Fig. 2.4) that the evaporation, E, from a clay

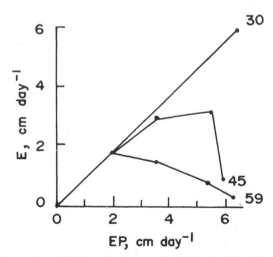

Fig. 2.4.   The evaporation rate, E, in a clay loam soil as a function of evaporativity, EP, at water table depth of 30, 45, and 59 cm, respectively (Schlesener, 1958).

loam soil was more closely related to the potential evaporation (evaporativity, *EP*) at shallower water table depths. At higher *EP*, the internal rate of water conductance, i.e., the hydraulic conductivity, itself becomes the limiting factor to meet the evaporation demand.

Keeping the above in view, Polynov as early as 1930 stated the concept of *critical water table depth*, which is defined as "the depth of water table below the soil surface above which the salts contained in the groundwater can rise through capillary action to the surface horizon to provoke soil salinisation under natural conditions".

Talsma (1963) observed a hyperbolic decrease in evaporation on lowering of water table and considered the depth of water table from which maximum evaporation rate was 0.1 cm/day, to be critical. Using this criterion, Khosla *et al.* (1980) observed 80 cm (Fig. 2.5) as critical water table depth for sandy loam soil of Sonepat district, Haryana, India, under bare soil conditions.

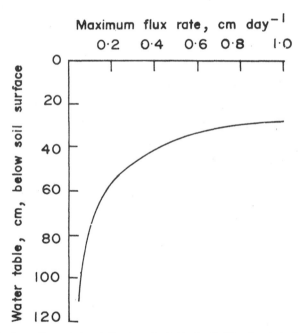

Fig. 2.5.   Maximum flux rate in relation to water table depth (Khosla *et al.*, 1980).

In general, a close relationship exists between the critical depth of groundwater and its salt content in order to cause salinisation of the surface soil (Fig. 2.6). The depth from which groundwaters are liable to salinise surface soils decreases with increase in their salt concentration. As a rough approximation for salt concentration of 1, 2.5 and 5 g, the critical water table depths are 3, 2 and 1 m, respectively.

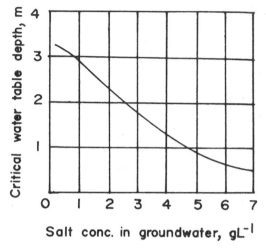

Fig. 2.6.   Relation between the salt content of groundwater and the critical depth of the water table.

Varallyay (1977) considered the maximum quantity of salts (QS) that can be accumulated in the soil profile during a year equal to the natural leaching potential of the soil and related it with maximum permissible (critical) flow velocity (Vm) as follows:

$$Vm = \frac{10 \cdot QS}{CwT} \qquad (3)$$

where Cw = the salinity of groundwater, g/l, and

T = time in days.

The depth of water table that prevented higher flux than *Vm* was defined as critical depth and was found to be 70 cm for sandy and 225 cm for loam soils.

2.3. Diagnosing Salinity Problems in the Field

In the field, saline soils can be identified by:
  i)   Presence of white crust of salts on the surface in a dry state (Fig. 2.7). This efflorescence may be wet, fluffy, or solid in consistency and light or dark in colour depending upon its main constituents. Abundance of $CaSO_4$ and $CaCO_3$ gives a fine, fluffy, dusty surface, while a mixture of $NaCl$ and Na_2SO_4 gives a crystalline white mass. $MgCl_2$ gives a dark crust that is highly hygroscopic.
  ii)   Good physical conditions and high permeability.

iii) High water table, mostly within 2 m of the surface. Subsoil water is mostly brackish and unfit for irrigation.

iv) Natural vegetation consisting of mainly small bushes and some halophytes such as *Cressa cretica, Cyperus rotundus, Chloris gayana, Sporobolus Pallidus, Dichanthium annulatum, Suaeda fruticosa,* and *Butea monosperma.*

v) In cultivated fields, patchy and stunted crop growth, often with a deep green to bluish colouration.

vi) Wilting sign of water stress in plants even when the soil apparently contains enough water.

vii) Generally a thin soil crust that may prevent emergence of seedlings of sensitive plants.

viii) Short life of buildings, farm machinery, and roads due to corrosive effect of salts.

2.4. ECe as a Measure of Soil Salinity

The soluble salt concentration of the soil is generally expressed in terms of electrical conductance of the saturation paste extract (ECe). In the laboratory saturation extract is obtained by suction or pressure from an equilibrated saturated soil paste. With increase in ECe, the soluble salts increase (Fig. 2.8). In general the concentration of soluble salts in me/l is 10 to 12.5 times the ECe, dS/m. The particular advantage of the saturation extract method for monitoring soil salinity lies in the fact that the saturation percentage (SP) is approximately twice the value of the water content at field capacity (FC). The soluble salt concentration in the saturation paste extract, therefore, tends to be one-half the concentration of the soil solution at the upper end of the field moisture range and about one-fourth the concentration that these soils would exhibit at the dry end of the field moisture range. The range of SP value commonly found in the soils of varying texture and the corresponding value of wilting point (WP) and field capacity (FC) are given in Table 2.2.

Table 2.2.   Moisture content (%) at WP, FC and saturation in soils of varying texture

Soil texture	Moisture percentage at		
	WP	FC	Saturation
Sandy loam	4	8	16
Silt loam	10	20	40
Clay	25	50	100
Peat	35	70	140

Fig. 2.7.   A layer of white fluffy salts present on the surface is a common characteristic of saline soils (Chhabra, unpublished).

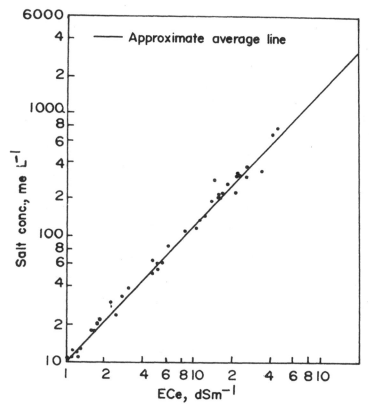

Fig. 2.8. Relation between EC of the saturation extract and the salt concentration (Richards, 1954).

The deleterious effect of high soluble salts on crop growth is mostly related to the increasing difficulty of extracting water from more concentrated solution. This is due to the increased osmotic pressure (OP), which is quantitatively related to the ECe of the soil.

The rootzone salinity is better characterised by the ECe as it is more relevant and takes into account the moisture differences due to texture of the soil. Figure 2.9 shows the relationship between ECe, OP, per cent salt, and crop response. The diagonal lines correlate ECe values with per cent salt content for various soil textures. For example, at ECe 4 dS/m and SP 75, the salt content of the soil is about 0.2 per cent. On the other hand, the same salt content in a sandy soil, say SP 25, would lead to ECe value in the SP extract of 12 dS/m. As the osmotic effects are more related to ECe of the soil, which is a direct measure of soil salinity, it is a better index of soil salinity than the per cent salt content.

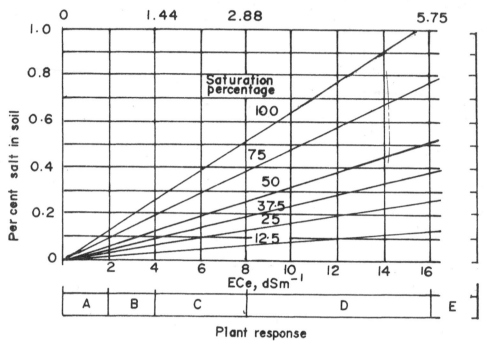

Fig. 2.9.   Relation between OP, ECe of saturation extract, per cent salt in the soil and crop response (Richards, 1954).

2.5. Monitoring Soil Salinity in the Field

Salinity measurement in the field is done with the help of porous-matrix salinity sensor, four-electrode units, and magnetic resonance techniques. For continuous monitoring of changes in soil salinity, over a period of several days or a leaching cycle, salinity sensors are the best (Richards, 1966). The salinity sensor measures the electrical conductivity of the soil water (EC_{sw}), which has been imbibed in a 1 mm thick ceramic disc between two screen electrodes.

Four-electrode units measure the EC of the bulk soil by measuring the resistance to current flow between a pair of electrodes inserted into the soil while an electric current is passed between a second pair of outer electrodes (Rhoades and Ingvalson, 1971). Since minerals are non-conductors, electric conduction is primarily through the soil solution and is a direct measure of its salt content. By changing the distance between the electrodes, salinity of a particular volume of soil or salinity distribution within a specified part of the rootzone can be monitored on a regular basis (Rhoades and van Schilfgaarde, 1976).

The dielectric constant of the medium as measured by the time domain reflectometry (TDR) is another method of monitoring soil moisture and its salt contents. The dielectric constant is determined by measuring the transit time of an electromagnetic pulse launched along a set of metallic parallel rods embedded in the soil. Salinity affects the attenuation of the electromagnetic pulse. By measuring the transit time and the degree of attenuation, this technique helps in measurement of soil salinity and water content independently with the same probe (Dalton *et al.*, 1984).

2.6. Effect of Salinity on Plant Growth

Excess amount of soluble salts affects the plant in the following ways:

2.6.1. Water Availability

With increase in salt concentration of the soil, the osmotic pressure of the soil solution increases (Fig. 2.10) and plants are not able to extract water as easily as they can from a relatively non-saline soil. Soluble salts exert this potential over and above the matrix potential already existing in the soil. Thus, as the salt concentration, i.e., ECe of the soil, increases the water becomes less available (Fig. 2.11) to the plant even though the soil may contain water and appears quite moist.

Osmotic pressure is a colligative property of the solution just like boiling point or freezing point and depends upon the number of particles present in the solution. Mathematically:

$$OP = cRT \tag{4}$$

where OP =   osmotic pressure in bars,

        c =   molar concentration,

        R =   gas constant, 0.082 L Atm/mole/°C, and

        T =   absolute temperature.

The osmotic pressure of the soil extract can be estimated by the empirical relationship:

$$OP = 0.36 \times ECe \; (dS/m) \tag{5}$$

At about 1.44 bars, which corresponds with ECe 4 dS/m, the plants start experiencing severe water stress due to physiological unavailability of water. Thus, in saline soils, in spite of the fact that water may be physically present, it becomes unavailable to the plants and this phenomenon is normally referred as "physiological drought".

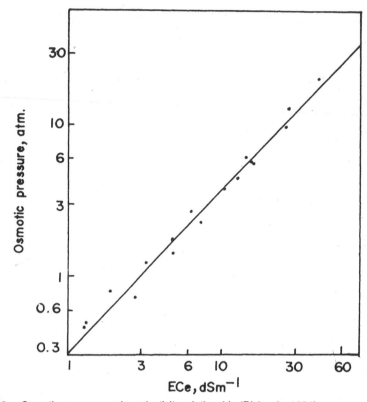

Fig. 2.10.   Osmotic pressure and conductivity relationship (Richards, 1954).

2.6.2. Effect of Salinity on Evapotranspiration

Evapotranspiration (ET) is a complex process influenced by soil, plant, and climatic factors including temperature, humidity, wind velocity, sunshine, and solar radiation. It has been shown by several workers that ET decreases with increasing salinity level in the growth medium. Hayward and Spurr (1944) observed that it is due to the difficulty the plants have in absorbing water from a salt solution whose osmotic pressure is higher than that of the cell sap. A direct relationship between the effect of salinity on ET and yield of Sudan grass (Fig. 2.12) was found by Balba and Soliman (1978), which showed that with increase in salinity ET decreased, as did the yield of the crop.

Increasing salinity reduces ET because of:
a) low water availability caused by high osmotic pressure of soil solution,
b) restricted root growth resulting in less water-absorbing area,
c) decreased leaf area, and
d) higher retention of water within the plant to dilute the absorbed salts.

Total soil water potential, bars

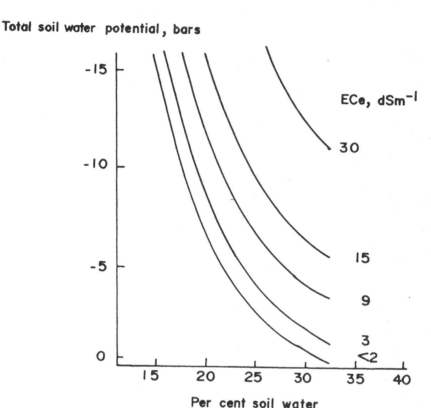

Fig. 2.11. Available water as affected by average soil salinity for a clay loam soil (Ayers and Westcot, 1976).

These factors reduce the water use efficiency, resulting in decreased plant growth and yield.

2.6.3. Specific Toxic Effect of Ions

Higher concentration of individual ions in the root environment or in the plant may prove toxic to the plant or may retard the uptake (absorption) and metabolism of the essential plant nutrients and thus affect the normal growth of the plant. Antagonism between Cl^- and $H_2PO_4^-$, Cl^- and NO_3^-, Cl^- and SO_4^{2-}, and Na^+ and K^+ may disturb the normal nutrition of plants. Sulphates that take part in the metabolic process as an integral part of proteins and enzymes may disturb the system and are normally more toxic than chlorides. However for certain fruit trees like citrus, grapes, and many woody trees, Cl^- is more toxic than SO_4^{2-} and causes characteristic "leaf burn" symptoms.

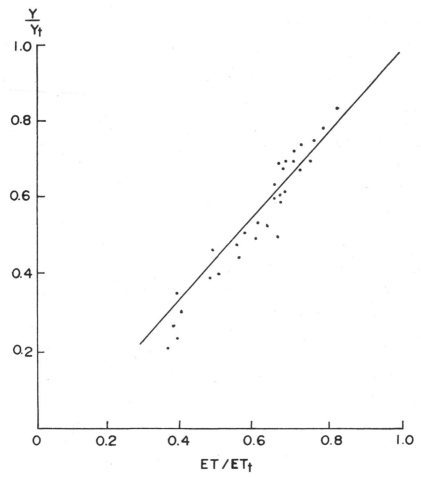

Fig. 2.12. Relationship between reduction in evapotranspiration and reduction in yield of Sudan grass due to salinity. Y_t and ET_t are yield and evapotranspiration with tap water, other Y and ET values were obtained by increasing salinity (Balba and Soliman, 1978).

Elements such as B, Li, and Se may also be present in toxic concentrations, especially if the underground water is saline, e.g., a boron concentration of more than 4 mg/l in saturation paste extract is toxic for most of the agricultural crops.

2.7. Effect of Waterlogging on Crop Growth

A shallow water table, which may result in waterlogging and accumulation of salts, is the most common cause of decreased productivity of irrigated agriculture. Waterlogging can cause changes in

— water and thermal regimes,
— soil strength,
— oxygen availability,
— accumulation of CO_2, HCO_3^-, CO_3^{2-}, and S^{2-},
— oxidation-reduction status of soil,
— soil pH,
— nature of microbes; from aerobic to facultative-anaerobic to anaerobic,
— nutrient availability by altering solubility, uptake, translocation, and interactions,

and can ultimately affect the crop yield. Soil becomes reduced and unfit to support plant growth if the redox potential is less than + 200 mV (Table 2.3). Drainage, surface or sub-surface, is a must not only to leach excess salts but also to keep the soil free from waterlogging.

Table 2.3.   Redox potential and the status of soil

Kind of soil	Redox potential, mV
Aerated, well drained	+ 700 to + 500
Moderately reduced	+ 400 to + 200
Reduced	+ 100 to − 100
Highly reduced	− 100 to − 300

2.8. Salinity Tolerance of Crops

Tolerance of crops to soil salinity depends upon the nature of plant, stage of growth, type of salinity, soil fertility, climatic factors, and the frequency of irrigation.

Climate is the single most important factor and plays a pivotal role in plant response to soil salinity. In humid areas plants may be able to tolerate higher salinity than in arid areas. Similarly, winter corn or winter sunflower is able to tolerate higher salinity than summer corn and summer sunflower.

Usually, it is difficult to fix a limit of salinity where the plant will fail to grow. Generally, the plants suffer a slow death with increase in soil salinity. However, based on average soil ECe and generalised plant response, the following salinity classes (Table 2.4) have been recognised:

2.8.1. Quantification of Salinity Effects

Numerous studies have been conducted on the relationship between plant growth and soil water salinity, quantitative expression for which would enable us to predict the effect of a given salinity level on yield.

Table 2.4. Soil salinity classes and crop growth

Soil salinity class	ECe, dS/m	Effect on crop plants
Non-saline	< 2	Salinity effects negligible.
Slightly saline	2–4	Yield of sensitive crops may be restricted.
Moderately saline	4–8	Yield of many crops restricted.
Strongly saline	8–16	Only tolerant crops yield satisfactorily.
Very strongly saline	> 16	Only a few very tolerant crops yield satisfactorily.

The USDA Salinity Laboratory Staff expressed the effect of salinity on plant production in terms of the salt concentration, which reduced the yield by 50 per cent as compared with non-saline conditions (Richards, 1954).

Salt tolerance data from experiments in many parts of the world were collected by Maas and Hoffman (1977), who used them to calculate the effect of salinity on the relative yield of several crops using the following equation:

$$Y = 100 - b \, (ECe - a) \tag{6}$$

where Y = relative crop yield (per cent),

ECe = salinity of the soil saturation extract, dS/m,

a = salinity threshold value (ECe for 100 per cent yield), and

b = yield loss per unit increase in salinity.

The ECe values other than those associated with 100 per cent yield were calculated from the yield equation of Maas and Hoffman (1977) as follows:

$$ECe = \frac{100 + ab - Y}{b} \tag{7}$$

Another way of expressing the yield salinity relationship is given by the following equation:

$$Y = \begin{bmatrix} Ym & 0 < C < Ct \\ Ym - Ym \cdot S \cdot (C - Ct) & Ct < C < Co \\ O & C > Co \end{bmatrix} \tag{8}$$

where Ym = yield under non-saline conditions, i.e., control yield,

C = average rootzone salinity during the growth season,

Ct = threshold concentration, i.e., salinity up to which there is no decrease in yield,

Co = salinity beyond which the yield is zero, and

S = absolute value of slope of the response curve between Ct and Co.

This relationship can now be solved as piecewise linear crop salt tolerance function as represented graphically in Fig. 2.13.

2.8.2. Relative Tolerance of Crops to Soil Salinity

Using the piecewise linear crop salt tolerance response function, Maas and Hoffman (1977) have summarised the salinity response of different crops (Fig.

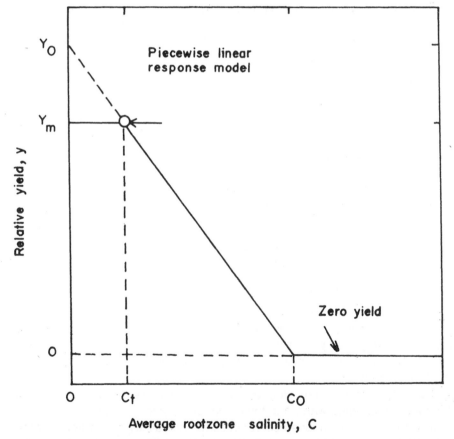

Fig. 2.13.   Graphical representation of the piecewise linear crop salt tolerance response function (Maas and Hoffman, 1977).

2.14 a–h). In general, beans are more sensitive, followed by fruit trees like citrus and grapes, while crops like barley and sugarbeet are more tolerant to soil salinity. Based on several field studies, their relative tolerance is listed in Table 2.5.

Table 2.5.   Relative tolerance of crops to soil salinity

	Sensitive ECe, 2–4 dS/m	Semi-tolerant ECe, 4–8 dS/m	Tolerant ECe, 8–16 dS/m
a)	**For inland saline soils**		
	Citrus spp.	Pomegranate	Date palm (*Phoenix dactylifera*)
	Cowpea (*Vigna sinensis*)	Wheat (*Triticum aestivum*)	Barley (*Hordeum vulgare*)
	Gram (*Cicer arietinum*)	Oats (*Avena sativa*)	Sugarbeet (*Beta vulgaris*)
	Peas (*Pisum saccharatum*)	Rice (*Oryza sativa*)	Spinach (*Spinacea oleracea*)
	Groundnut (*Arachis hypogaea*)	Sorghum (*Sorghum bicolor*)	Rape (*Brassica* spp.)
	Lentil (*Lens esculenta*)	Maize (*Zea mays*)	Cotton (*Gossypium hirsutum*)
	Mung (*Phaseolus aureus*)	Sunflower (*Helianthus annuus*)	
		Potato (*Solanum tuberosum*)	
b)	**For coastal saline soils**		
		Barley (*Hordeum vulgare*)	Rice (*Oryza sativa*)
		Linseed (*Linum usitatissimum*)	
		Raya (*Brassica juncea*)	
		Safflower (*Carthamus tinctovillis*)	
		Chillies (*Capsicum annum*)	

Source: Compiled from several studies.

For a given crop, the relative tolerance to salinity may vary greatly from variety to variety. Few promising varieties of rice suited for coastal saline soils are listed in Table 2.6. Promising varieties of other crops suited for inland saline soils are listed in Table 2.7.

Rice crop is most suitable for coastal areas where yield is affected both by the depth of standing water and the extent of its salinity. Table 2.6 gives the relative tolerance of various rice varieties to soil salinity and water submergence depth. The relative tolerance of varieties of different crops suitable for inland saline soils is given in Table 2.7.

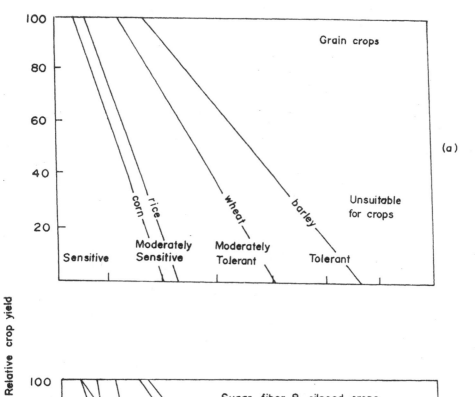

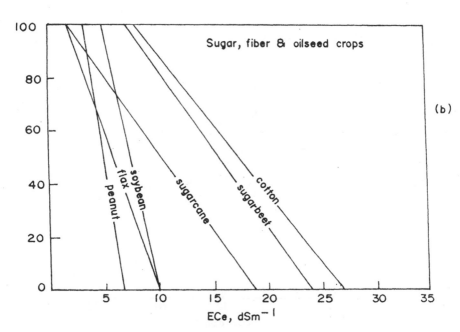

Fig. 2.14.   Relative tolerance of crops to salinity (Maas and Hoffman, 1977).

(Contd.)

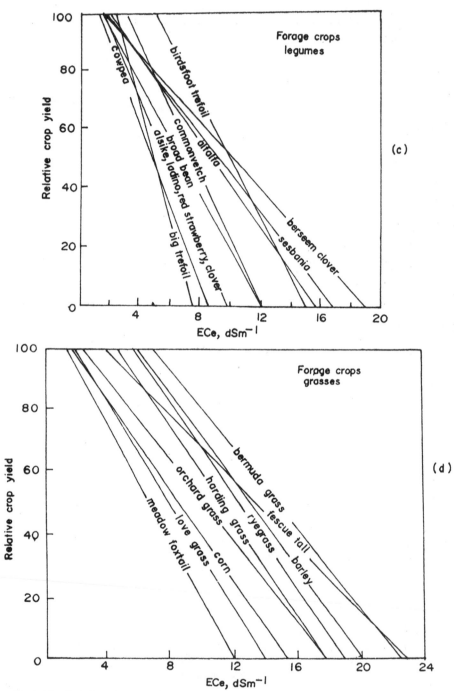

Fig. 2.14.   Relative tolerance of crops to salinity (Maas and Hoffman, 1977).

(Contd.)

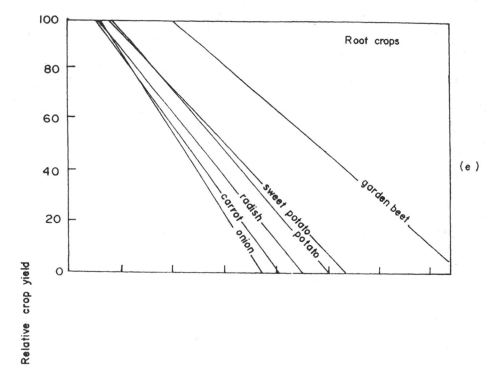

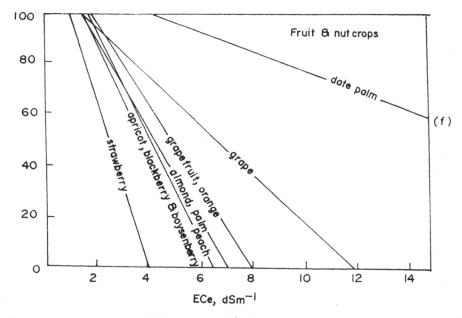

Fig. 2.14.   Relative tolerance of crops to salinity (Maas and Hoffman, 1977).

(Contd.)

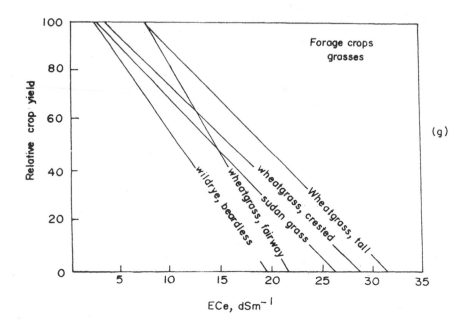

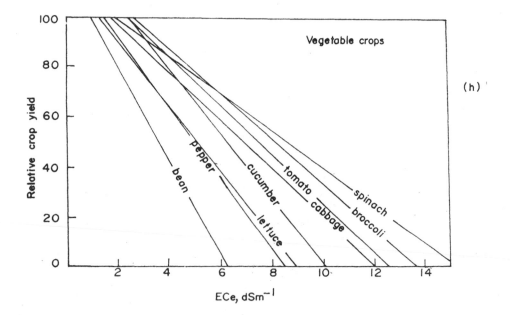

Fig. 2.14.   Relative tolerance of crops to salinity (Maas and Hoffman, 1977).

Table 2.6. Relative tolerance of rice varieties to soil salinity and water submergence depth in coastal saline soils, Canning Town, India

Salinity level ECe, dS/m	Depth of water submergence		
	≤ 20 cm	20–30 cm	30–50 cm
Low < 5	CSR-4, Jaya RP6-510, IET-7904	Jaya, IET 7904 Mahsuri, Pankaj NC-1281	SR 26-B, NC-1281 Kalamota, IET-6905
Moderate 5–8	CSR-1, 2, 3, 4 IET-7904, 7905 7907, 9341	CSR-1, 2, 3, 7 IET-7904,7905 Nonabokra, Matla Hamilton	SR26-B, CSR-6, Hamilton, Matla, NC-540, Kalamota
High 8–10	CSR-1, 3, 4 Velki	Velki, Bokra CSR-1, 2, 3, 7, Nonabokra, Matla, Hamilton	

Table 2.7. Varieties of different crops suitable for inland saline soils

Crop	Varieties
Rice	CSR 1, 2, 3, 5, 81-H, M21-2-4, 57-7, 3-11, 3-12, 3-13, RR-106, Jaya
Wheat	KRL 2-10, KRL 3-4, KRL 4-4, WH 157, HD-1982, Sonalika, PBW-65
Barley	CSR-1, CSB-2, 3, CS-54, CS 80-2, DL-2000, DL-348, DL-88, DL-352, Ratna, P-469, Jyoti
Pearl millet	HV-82, HC-5, WCC-75, ICMS-8010, ICMV-81111, KBP-3
Raya	RID-142, RLM-608, Varuna
Sugarbeet	Ramolave, Ramonskaya, Maribo Marinapoly, Polyrave

2.9. Mechanism of Salt Tolerance in Plants

Many plants have developed special methods to avoid or tolerate salinity and thus grow in the hostile environment. Some of these are:

a) Salt exclusion or avoidance
The roots of many plants are capable of avoiding the absorption of salts not required by the growing plants. This is achieved through selective ion absorption by the cell membrane. Many of the halophytes growing in the sea may adopt this technique. However, normal agricultural plants cannot avoid the absorption of salts, especially when their concentration increases above a certain limit, usually above ECe 10 dS/m.

b) Immobilisation within the plant
Many of the plants just try to restrict the absorbed toxic salts in the root or lower leaves and avoid their translocation to the growing active leaves and the

reproductive parts, where they can do the most harm. Normally, the concentration of toxic elements such as Na is less in the seed and at a maximum in older non-active leaves. Some of the plants (*Atriplex* species) immobilise the absorbed salts by storing them in special structures such as salt glands and hairs and thereby avoid their toxic effects.

c) Excretion

Some plants, such as cauliflower and tomato, are able to maintain their salt balance by excreting the absorbed salts and thus throwing them away from their system. This is achieved through specialised structures like hydathodes, which excrete salts through the process of guttation.

Using ^{36}Cl, Chhabra *et al.* (1977) observed that tomato plants are able to excrete large quantities of salts through hydathodes. The autoradiographs (Fig. 2.15) showed that ^{36}Cl accumulated in the margins of the young leaves, from where it was excreted out. One can actually see the saltish guttation fluid hanging in the form of droplets on the margin of leaves of many plants early in the morning.

Plants growing in sea water have a special Na-pump located in the cell membrane. This keeps throwing out Na from the inner side to the outer side of the cell and thus keeps the concentration of this toxic element within the cell under control.

d) Adjustment in osmotic pressure

To match the osmotic pressure of soil solution and thus be able to extract water, spinach, sugarbeet, tomato, and other plants adjust to soil salinity by increasing their internal osmotic pressure. For this they use the absorbed inorganic salts. Most of the inorganic salts are stored in vacuoles of the cell so as not to interfere with the biological activities carried out in the cytoplasm.

In many plants, this internal adjustment in osmotic pressure is made through organic metabolites like sugars, alcohol, and amino acids. Because of this, fruits of plants grown under stress conditions are many times sweeter than those grown under stress-free conditions.

2.10. Criteria for Selection of Salt-tolerant Crops

Depending upon the objective, the following criteria are generally used to screen crops or varieties for their salt tolerance:

a) Germination: duration, percentage
b) Survival
c) Root growth
d) Growth rate, height, biomass production
e) Leaf injury, per cent chlorophyll content
f) Na and Cl accumulation

Fig. 2.15. Autoradiographs showing accumulation of ^{36}Cl in the margins of tomato leaves, from where these are excreted out through hydathodes (Chhabra *et al.*, 1977).

g) Na/K or Ca/K ratio
h) Ethylene production, proline concentration
i) Concentration of nutrients in active tissue
j) Leaf to leaf compartmentation

k) Yield—Numbers of tillers, ear length
   1,000 grain weight
   Total yield
   Grain yield
   Grain/straw ratio (harvesting index)

2.11. Methods of Reclaiming Saline Soils

Unproductive saline soils can be made fit for agricultural production First, the soluble salts must be removed from the rootzone and the source of salts cut to prevent resalinisation of the soils. Some of the commonly adopted methods to achieve this are:

a) *Scraping:* The salts that have accumulated on the surface can be removed by mechanical means. This is the simplest and most economical way to reclaim saline soils if the area is very small, e.g., small garden lawn or a patch in a field. Scraping of the salts improves plant growth only temporarily, as the salts accumulate again and again.

b) *Flushing:* Sometimes washing away the surface salts by flushing water over the surface is helpful in removing the salts. This is especially practicable for soils having a crust and low permeability. However, this method is also not very sound as the salts keep accumulating on the surface again. Moreover, if a safe disposal is not available, the salt-rich water cannot be disposed of without creating problems in other areas.

c) *Leaching:* Leaching with water, irrigation or rain, is the only practical way to remove excess salts from the soil. It is effective if drainage facilities are available, as this will lower the water table and remove the salts by draining the salt-rich effluent. This aspect will be discussed in Chapter 8, "Irrigation and Salinity Control".

2.12. Management of Saline Soils

For management purposes the following four categories of saline soils have been identified:

a) Coastal and deltic saline soils.
b) Inland saline soils with shallow water table and brackish groundwater.
c) Good soils with deep water table but brackish underground water as the only source of irrigation.
d) Acid-sulphate soils.

2.12.1. Coastal and Deltic Saline Soils

Management of these soils centres around preventing the ingress of sea water through high tides, rise in water table, and back flow into the rivers and estuaries. High and structurally stable earthen dykes should be built to prevent the entry of sea water into the cropped area. In addition to this, back flow of sea water, during the off-rainy season, in the rivers should be controlled with the help of flap gates in such a way that it does not salinise the groundwater and raise the salinity of the soil. This will cut the source of salts, i.e., sea water and thus prevent the salinisation of the area. In these areas, groundwater should be exploited carefully and only the good quality rain water (floating over the brackish water) should be skimmed from the shallow wells. Over-exploitation of groundwater will encourage ingress of sea water through the subsoil and result in salinisation of the area.

2.12.2. Inland Saline Soils

Removing of salts by leaching with good quality water and lowering of water table through surface and sub-surface drainage are the best possible ways to reclaim saline soils. But in areas where only saline irrigation water is available or when shallow saline water table prevails and soil permeability is low, achieving non-saline conditions may not be practically and economically feasible. Under such conditions, one has to live with the salinity and minimise or modify its adverse effects by adopting various cultural practices.

a) Choice of crop and cropping pattern
One strategy available to farmers with saline soils is to select crops or varieties that can tolerate moderate to high soil salinity. An appropriate crop rotation, which should include crops with low evapotranspiration rate and different rooting pattern, can help to maintain the productivity of saline soils. Except for coastal saline soils with shallow water table and acid-sulphate soils, rice should never be grown in saline soils, as it will contribute to rise in water table and salinity of the soil.

b) Proper seed placement
Plants are generally most sensitive to soil salinity during the germination and early vegetative stages of growth and become more tolerant at later stages (except in rice, where salinity stress causes maximum damage at pollination stage and results in failure of grain formation). Because of greater sensitivity at seedling emergence stage, it is important to keep salinity levels in the seed bed as low as possible at planting time. If salinity levels reduce plant stand, potential yields may decrease far more than those predicted by salt tolerance data.

Low salinity in the rootzone can be achieved by manipulating the soil surface conditions, i.e., bed shape and irrigation management.

It is well known that salts tend to accumulate on the ridges away from the wet zone when furrow irrigation is adopted (Fig. 2.16). Putting the seed on

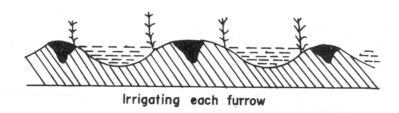

Irrigating each furrow

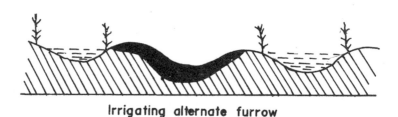

Irrigating alternate furrow

Fig. 2.16.   Pattern of salt accumulation in furrow irrigation and methods to avoid its toxic effect by proper placement of seed. Dark patches indicate to the zone of salt accumulation (Oster *et al.*, 1984).

off-centre slope of the single row will put the seed in minimum salinity and optimum moisture conditions. Under high salinity the alternate row should be left unirrigated, this will ensure maximum accumulation of salts in that (unir-rigated) area and leave the slopes of irrigated furrows free of salts and fit for planting seeds.

If salinity is due to brackish water and good quality water is available only in limited quantities, then the best alternative is to apply pre-sowing irrigation with good quality water. This will ensure minimum salinity in the soil at the seed germination stage.

c) Method of raising plants
Germinating seeds are severely affected by the osmotic effects of salinity, resulting in higher mortality and thus poor crop stand. So, wherever possible (e.g., in vegetables, flowers, fruit trees), the crop should be raised by transplanting seedlings. For this, strong, older seedlings raised in nursery on good soil should be used.

d) Method of water application
The method of irrigation affects the depth of irrigation, runoff, deep percolation losses, uniformity of application, and thereby salinity of the soil.

Under furrow or drip irrigation (Fig. 2.17), salinity levels are low immediately beneath the water source and increase with depth. Midway between the furrow or drip sources, soil salinity is medium (levels being highest at the surface) and moisture optimum for plant growth.

Sub-surface irrigation systems provide no means of leaching the soil above the source of water. Continuous upward water movement to the evaporation surface causes salts to accumulate near the soil surface. Unless the soil is leached by rainfall or surface irrigation, salt will certainly accumulate to toxic levels.

Sprinklers often allow much efficient use of water and a reduction in deep percolation losses. Under sprinkler irrigation, lateral salt distribution is relatively uniform, but soil salinity increases with depth. It is very effective in leaching

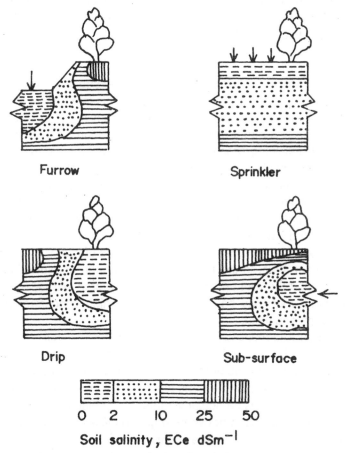

Furrow   Sprinkler

Drip   Sub-surface

0   2   10   25   50

Soil salinity, ECe dSm^{-1}

Fig. 2.17.   Zone of salt accumulation under different systems of irrigation (Bernstein and Francois, 1973).

the surface soil and provides a non-saline environment free of crust for germination and initial stages of plant growth. It has been quite intensively used for growing lettuce.

However, if the quality of water is bad, then the sprinkler-irrigated crops are subject to damage by both soil salinity and salt spray to the foliage. Salts may be directly absorbed by the leaves, resulting in leaf injury (necrosis on the margins) and loss of leaves. In crops that normally restrict salt movement from the roots to the shoots (e.g., citrus, grapes, tree species), foliar salt absorption can cause serious problems not normally encountered with surface irrigation system. For example, water with EC 4.5 dS/m when sprinkler-applied reduced the yield of peppers by about 50 per cent, but by only 16 per cent when applied to soil surface. Similarly, damage was greater to the cotton plant than to wheat when saline water was applied through sprinkler irrigation. Damage to leaves is minimal at low temperatures and under high humidity. That is why sprinkler irrigation should be applied at night.

e) Frequency of irrigation

If the crop depletes the soil water and suffers water stress between irrigations, one obvious solution is to irrigate more often. This is a simple and effective solution particularly for shallow-rooted crops or on soils whose initial infiltration rate is high but drops rather quickly (due to use of high SAR or RSC waters).

Immediately following irrigation, the soil water content is at a maximum and concentration of dissolved salts is at a minimum (Fig. 2.18) but each changes, as the water is consumed. As the soil available water is depleted (by evaporation or plant uptake), the water deficit and osmotic effects become greater during this period. The resulting increase in salt concentration can add appreciably to the severity of the salinity problem.

This may become very important during periods of high ET demand when water movement towards the root is not fast enough to dilute the salt concentration around the root and also to supply the crop with adequate water. Thus, the plant root may be exposed to very high concentration of salts affecting growth and yield.

More frequent irrigations can maintain better water availability and decrease the salinity to which the crop is exposed. With regard to permeability, this will also maintain a lower soil sodium adsorption ratio, since dilution favours the adsorption of Ca and Mg over Na and losses of Ca due to precipitation will be kept to a minimum. This is particularly important for high CO_3^{2-} and high SAR waters where severe drying between irrigation cycles is believed to remove appreciable quantities of Ca by precipitation.

However, care should be taken not to use too much water to achieve frequent irrigations. This will lead to appreciable increase in water use and

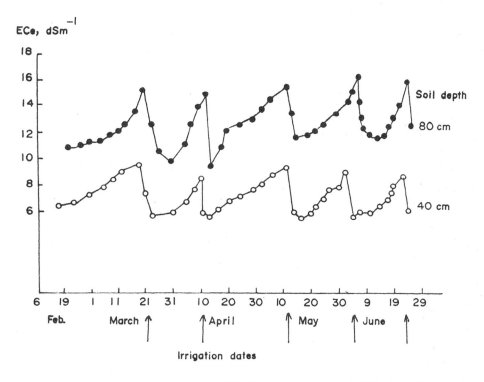

Fig. 2.18. Changes in salinity of soil water (ECe) between irrigations of alfalfa due to the use of stored water and irrigation (Ayers and Westcot, 1976).

thus higher salts and more drainage water. So an efficient method such as the sprinkler will be more appropriate to be used with more frequent irrigations.

f) Rootstock-scion relationship
Unlike most annual crops, fruit trees and other woody species are more sensitive to specific salt constituents and often develop leaf burn symptoms from toxic levels of Na^+ and Cl^-. Many wild rootstocks are available that can tolerate high salinity and do not transport these toxic ions to the shoots. But their fruit-bearing capacity and/or quality of fruit is very poor. These rootstocks can be exploited for grafting a good quality but salinity-sensitive scion. This is already being done in fruit trees like mango, citrus, and guava and ornamental plants like rose.

g) Use of mulches
During the early stage of crop growth much of the water evaporates from the surface, leaving behind the salts, which affect the germination by their direct

toxicity, osmotic effects, and/or development of hard crust on the surface. Evaporation of water and its consequences can be checked by providing a mulch on the surface. For this, crop residues like straw or a thin plastic sheet or even the natural soil mulch can be used. The mulch proves more beneficial in winter when in addition to conserving the moisture, and preventing the build-up of salts, it also maintains proper temperature of the soil, and thereby helps in early germination of the seeds.

h) Proper use of plant nutrients

Nutritional disorders can accentuate the yield limitations imposed by the osmotic effects of salinity. High water table and poor aeration can limit the root length and its absorbing area and this may restrict its ability to absorb plant nutrient in required amounts. Even the metabolism of the absorbed nutrient into organic metabolites can be adversely affected by salinity.

The harmful effect of moderate salinity can be alleviated by the judicious use of fertilisers. Specific toxic effects of Cl^- and SO_4^{2-} can be minimised by proper application of P- and N-fertilisers. Many times, leaching losses of nutrients can be very high and necessitate the application of fertilisers in adequate amounts.

Quite often it is believed that addition of inorganic fertilisers may aggravate the problem caused by excess salts. However, considering the amount and nature of fertiliser salts, that seems to be a very remote possibility. In practice, lack of essential nutrients may in fact be responsible for reduced yields in many saline soils.

2.12.3. Acid-sulphate Soils

Acid-sulphate soils are characterised by low pH, ranging from 3.5 to 4.0, and medium to high salinity. These soils are mostly met with in coastal areas where decomposition of high organic matter (originating mostly from pollustric vegetation like mangroves or reed swamps) under anaerobic conditions in the presence of brackish sea water results in the formation of pyrite (FeS_2) (Pons, 1973). A slow sedimentation rate to allow sufficient time for reduction of iron and tidal flushing to remove HCO_3^- ions is also necessary for the formation of these soils. The pyritic layer occurs at a depth of about 50 cm and on oxidation releases H_2SO_4, lowering the pH of soil and creating Fe and Al toxicity and associated nutritional disorders. Such soils exist in Sri Lanka, Indonesia, Thailand, Vietnam, India, Senegal, and other countries and pose special problems for their reclamation and management. Composition of saturation extraction of a typical acid-sulphate soil is given in Table 2.8.

Plant growth in these soils is adversely affected by

(a) low pH,

(b) high soluble salts,

Fig. 2.19. Acid-sulphate soil showing the presence of pyritic layer (Chhabra, unpublished).

Table 2.8. Characteristics of an acid-sulphate soil from Calicut, Kerala, India

Depth, cm	pH_s	ECe, dS/m	Composition of saturation extract, me/l				Clay
			Na^+	$Ca^{2+} + Mg^{2+}$	Cl^-	SO_4^{2-}	%
0–9	4.4	43.6	322	123	354	120	48
9–20	4.8	12.6	100	38	106	36	39
20–36	4.8	11.7	82	41	86	53	46
36–70	4.5	8.4	76	25	57	65	63
> 70	3.4	38.6	285	112	337	75	50

(c) toxic concentration of Fe and Al and low availability of Ca and P,

(d) shallow water table resulting in waterlogging, and

(e) low base saturation.

Reclamation of these "wet desert soils" poses a special constraint. When these soils are drained, FeS_2 present in them (Fig. 2.19) gets oxidised, releasing free H_2SO_4, which further lowers the pH. These soils, therefore, should not be drained at all. Since these soils are to be managed as water-logged soils, lowland rice is the only suitable choice. For getting optimum yield of rice, it is important to neutralise excess acidity and to minimise toxic effects of Fe and Al by raising the pH, for which liming is essential. Considering their heavy texture, normally 5 to 10 t/ha of lime is needed to raise the pH and bring it between 5.5 and 6.0. Repeated applications of lime, after every 2 to 3 years, are required to neutralise acidity that is regularly being produced because of the presence of FeS_2.

In the temperate regions where rainfall is high and acidic drainage water can be dumped into the sea, initial leaching can be done with sea water, as was done in The Netherlands. This neutralises acidity, removes Fe and Al toxicity, and eliminates the need to lime the soil. The final leaching is done with rain water and mostly the area is developed and managed as pasture land.

In addition to raising pH and applying nitrogen, these soils require high amounts of P fertilisation. Rock phosphate, which contains water-insoluble P, is the most economical fertiliser for such soils as it supplies both phosphorus and calcium. Moreover, to get best results, those varieties of rice that are tolerant to Fe and Al toxicity and are efficient P absorbers should be raised on these soils.

For raising other crops, an intensive shallow drainage system commonly known as "Serjan" system of farming as practised in Indonesia is best suited for acid-sulphate and peat soils. In this system, raised beds are formed with alternate rows of shallow ditches or trenches dug at narrow spacing. The soil of the raised bed remains drained and leached of the salts and excess acid by rain water. So this area is used for growing crops like corn, cassava, groundnut, and beans, while the low-lying area is used for growing rice. Such a system is highly suitable for areas having high rainfall uniformly distributed throughout the year.

2.13. Rice in Saline Soils and Soils Irrigated with Saline or Sodic Waters

Rice is not the most tolerant crop to excess salinity, but its cultivation is favoured in saline soils, especially in the initial stages of reclamation of waterlogged saline soils. This is due to the fact that:

a) no other crop can tolerate excess water or poor aeration in the rootzone as efficiently as rice, and

b) a system of lowland rice culture that requires maintenance of standing water throughout the growing season helps in leaching and dilution of salts and thus lowers the effective rootzone salinity.

In practice, lowland rice at no stage is subjected to high salinity stress as otherwise indicated by the initial analysis made on dry soil samples. van Alphen (1975) reported that though salinity may be very high initially, it is significantly reduced (Fig. 2.20), especially in the upper few centimetres, after one or two rains, enabling transplanting of the seedling in a relatively good soil. Such an advantage is not available to other crops grown under arable conditions because the salinity stress as measured by initial soil analysis is more or less indicative of the effective rootzone salinity.

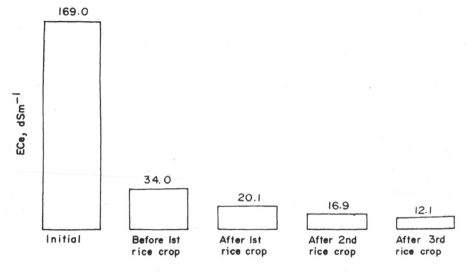

Fig. 2.20. Changes in upper 10 cm of a highly saline soil as affected by growing of rice crop (Adapted from van Alphen, 1975).

Moreover, growing of paddy while salts are leached with good quality water helps in raising a crop and thus economises on the cost of leaching. However, under such conditions, which correspond to leaching under continuous ponding of irrigation water, the efficiency of leaching will be less than that of the intermittent ponding.

Rice should never be grown on inland saline areas because, due to excessive use of water as conditioned by its cultural practices, it may result in rise in water table and be a cause of salinity development for the crops following rice. Similarly, where use of high RSC waters result in poor permeability and thus help in better maintenance of waterlogged conditions, the farmers tend to take rice crops. There also its cultivation should not be encouraged as the development of sodicity due to excessive use of high RSC water during the rice growing season may be detrimental to the following crops. Introduction of rice will thus decrease the total productivity of the area.

3

Alkali Soils

Alkali soils, also known as sodic or solonetz soils, are those that contain measurable amounts of soluble salts capable of alkaline hydrolysis, mostly CO_3^{2-} and HCO_3^- of Na^+, which in the presence of $CaCO_3$ give the soils high pH and poor physical conditions. These soils have pHs > 8.2 and ESP >15. The electrical conductivity of alkali soils is variable but normally less than 4 dS/m. Neutral salts like NaCl and Na_2SO_4 may also be present, but in small quantities. A sparingly soluble salt like gypsum, which is present in nearly all saline soils, is always absent in alkali soils. Because of high pH and Na_2CO_3, a part of the organic matter gets dissolved and imparts a dark black or brown colour to both the soil and its aqueous extract. For this reason these soils are also commonly known as black alkali soils.

3.1. Characteristics of Soil Profile and Saturation Extract

The top layer of these soils is highly dispersed and characterised by low permeability. These soils contain free $CaCO_3$ from 0.5 to 40 per cent. Many times, due to migration of clay and deposition of $CaCO_3$, they may contain a hard impervious layer (Fig. 3.1) at a depth of 30 to 100 cm. This layer acts as a physical impediment to the movement of water, leaching of salts, and vertical root growth. They may be without (Fig. 3.2) or with structural B horizon (Fig. 3.3). Exchangeable sodium percentage of the whole profile is generally very high.

In contrast to saline soils, the saturation extract of alkali soils (Tables 3.1, 3.2) contains measurable amounts of CO_3^{2-} and HCO_3^-; Cl^- and SO_4^{2-} may also be present in appreciable amounts. Among the cations, sodium is the dominant one, followed by Mg^{2+}; the concentration of Ca^{2+} is minimum, sometimes even in traces.

Fig. 3.1.   Typical soil profile of an alluvial alkali soil showing impervious CaCO$_3$ layer at a depth of 1 m (Chhabra, unpublished).

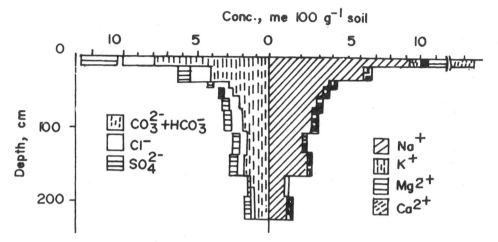

Fig. 3.2.   Salt profile of an alkali soil without a structural B horizon (Hungarian Danube Valley); (Szabolcs, 1974).

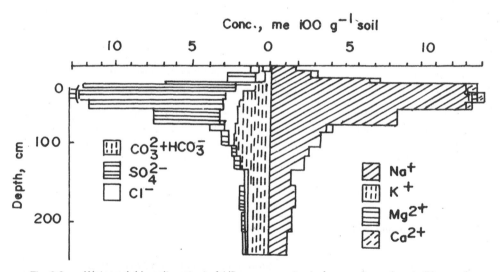

Fig. 3.3.   Water-soluble salt content of 1/5 aqueous extract of a meadow solonetz (Hungary) with a structural B horizon (Szabolcs, 1974).

3.2. Mode of Formation

Alkali soils can be formed under the following conditions:

a) Sodication due to repeated cycles of wetting and drying
Micro-relief of the area and geochemical mobility of the salts play a major role in the formation and distribution of alkali soils. These soils are found to be

Table 3.1. Characteristics of an alkali soil

Horizon	Depth, cm.	OM, %	CaCO₃, %	Sand	Silt	Clay	K, cm/ day	CEC, me/100 g	ESP
					%				
A3	0–20	0.8	4.7	22.1	32.2	43.7	nil	50.8	52.7
B21	20–67	0.5	7.4	21.7	23.1	52.6	nil	55.4	81.5
C1Ca	67–95	0.1	18.3	33.2	24.6	40.5	nil	37.4	89.4
C2Ca	95–100	nil	22.6	32.4	31.6	36.8	nil	33.6	92.3

Analysis of saturation paste extract

Horizon	pHs	ECe, dS/m	Ca^{2+}	Mg^{2+}	Na^+	CO_3^{2-}	HCO_3^-	Cl^-	SO_4^{2-}
					me/l				
A3	8.9	4.7	tr.	1.0	43.8	1.6	8.8	21.8	15.6
B21	9.8	4.1	tr.	2.0	45.7	2.0	12.3	17.5	16.2
C1Ca	10.0	4.7	tr.	1.5	48.6	2.5	13.3	22.0	12.7
C2Ca	9.7	3.6	tr.	2.0	27.9	2.5	8.5	14.5	8.4

Source: Chhabra (unpublished).

Table 3.2. Characteristics of an alkali soil near Lake Chad, Republic of Chad

Depth, cm	pHs	ECe, dS/m	Composition of 1:2 extract, me/100 g soil						
			Na^+	$Ca^{2+}+Mg^{2+}$	K^+	CO_3^{2-}	HCO_3^-	Cl^-	SO_4^{2-}
0–5	10.0	17.5	48	tr	2.6	6.0	35	2.7	2.6
5–10	9.7	12.5	37	tr	2.0	5.2	26	2.4	3.3
10–15	8.8	5.9	13	1.3	0.9	0.6	8	1.0	4.8
15–25	8.3	5.3	10	1.1	0.7	0.1	6	1.5	4.0
25–35	7.6	2.9	5	0.8	0.3	0	3	1.1	1.6
35–50	7.4	2.6	3	1.5	0.4	0	2	0.6	2.4
50–65	7.6	3.6	3	4.6	0.5	0	2	0.2	5.6

Source: C. Cheverry, quoted by Szabolcs (1979).

concentrated in the zone of mean rainfall 500 to 700 mm and occupy concave slope positions that are about 15–30 cm lower than the adjoining areas in otherwise flat terrain (Abrol and Bhumbla, 1971; Bhargava *et al.*, 1980).

It is observed that constant weathering of aluminosilicate minerals in the catchment area produces a steady supply of soluble salts comprising Cl^-, SO_4^{2-}, HCO_3^-, CO_3^{2-} of sodium, potassium, calcium, and magnesium, which migrate with the surface and subterranean waters and accumulate in the micro-depressions and the undrained areas during the rainy season. Due to the arid and semi-arid climate, the water evaporates in the post-rainy months leaving the salts behind. Repeated cycles of accumulation followed by drying favour the precipitation of soluble Ca as $CaCO_3$ (due to its low solubility) and excess accumulation of $NaHCO_3$ and Na_2CO_3, which in turn results in replacement of exchangeable Ca^{2+} by Na^+ and its subsequent precipitation. This process, known as sodication or sodiumisation, results in an increase of Na^+ concentration in the soil solution and the exchange complex (with simultaneous increase in pH) at the cost of Ca^{2+}, which precipitates as $CaCO_3$.

The sequential events leading to the formation of alkali soils are given below:

$$\text{Na-bearing primary minerals} + H_2O \xrightarrow{\text{weathering}} \underset{\text{clay}}{\text{H-silicate}} + \underset{\text{(transitory)}}{NaOH}$$

$$\underset{\substack{\text{(atmospheric or} \\ \text{biological)}}}{NaOH + CO_2} \longrightarrow NaHCO_3$$

$$2NaHCO_3 \longrightarrow Na_2CO_3 + CO_2 + H_2O$$

$$\underset{\text{(normal soil)}}{Ca^{2+}\text{-soil} + 2\,NaHCO_3} \longrightarrow \underset{\text{(alkali soil)}}{2Na^+\text{-soil} + Ca(HCO_3)_2}$$

$$Ca(HCO_3)_2 \xrightarrow{\text{Drying}} \underset{\text{(insoluble)}}{CaCO_3{\downarrow} + CO_2{\uparrow} + H_2O}$$

In an arid environment, alkali soils can also develop, as in close basins, where inflow waters may have a positive residual sodicity (Beek and Breemen, 1973).

Since $NaHCO_3$ and Na_2CO_3 salts accumulate mainly on the surface and their concentrations decrease with increase in depth, it is believed that alkali soils of the Indo-Gangetic alluvial plains are formed *in situ* by the direct process of sodication starting at the surface. This is further supported by the fact that the ESP of these soils is highest at the surface and decreases with increase in depth; and the underground waters of these regions are of low total salinity and mild alkalinity. This process of sodication may also affect the transformation of clay minerals. Pal and Bhargava (1979) reported that highly sodic environment of the soils results in degradation of clay minerals, as indicated by the trends of variations in molar ratio of amorphous SiO_2 and Al_2O_3. Degradation in these soils has been greatest in the surface horizon, decreasing with depth, and is in agreement with the degree of sodium saturation of the zone. In vertisols, sodication may lead to formation of chlorite from montmorillonite.

b) Sodication due to shallow brackish groundwaters
The alkali soils in Europe, the USSR, and Hungary (Kovda, 1973) and in many parts of the United States of America have been mostly evolved under the influence of shallow groundwater containing high residual sodium carbonate. These soils have a good A horizon but a natric (sodic) subsurface B horizon where the sodication has taken place due to shallow groundwater, low in divalent cations like Ca^{2+}

c) Sodication due to the use of high RSC waters
Irrigation with underground water containing high amounts of CO_3^{2-} and HCO_3^- is one of the chief contributing factors in the formation of alkali soils in many parts of the world. El-Elgabaly (1971) reported that this was the main reason for formation of alkali soils in Egypt. In most of the vertisol regions of India, alkalinity is associated with the practice of using high RSC waters for irrigation.

d) Sodication due to the use of saline irrigation waters low in Cl:SO₄ ratio
In areas of low rainfall, < 300 mm, sodication is also favoured by the continuous use of saline underground waters having a low Cl:SO₄ ratio. High SO_4^{2-} content of irrigation water lowers the concentration of Ca in soil solution by forming sparingly soluble $CaSO_4$ and this raises the SAR of the soil solution and the exchangeable sodium (ES) of the soil.
   Accumulation of salts in soils is a complex process that often defies interpretation in terms of simple consequence of climate and soil factors. Generally, a combination of more than one factor is required for their formation. A schematic diagram (Fig. 3.4) illustrates the role of elevation, annual rainfall, hydrological conditions, and geochemical mobility of salts in the formation of various soils.

3.3. Diagnosing Alkali Soils in the Field

In the field, alkali soils can be diagnosed by the following characteristics:
   a)   Soils remain devoid of any natural vegetation. Some very hardy grasses such as *Sporobolus marginatus, Leptochloa fusca, Chloris gayana, Cynodon dactylon, Sueda maritma,* and *Demostachia bipinnata* grow there.
   b)   When wet, alkali soils turn black because of the humic acid fraction of OM, which is dissolved by Na_2CO_3 at high pH.
   c)   They develop a pink colour (Fig. 3.5) when phenothphalene indicator is applied to them.
   d)   They are very slippery and soft when wet but very hard when dry.
   e)   Upon drying, deep cracks, 1–2 cm wide, develop in the soil (Fig. 3.6) which close when wetted.

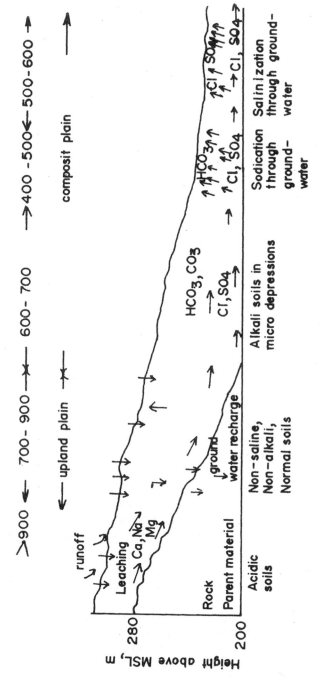

Fig. 3.4. Schematic distribution of salt-affected soils as influenced by precipitation, topography, hydrological, and hydrochemical factors in Haryana state, India (adapted from Bhargava *et al.*, 1980).

f)   The surface soil develops a hard crust with typical convex surfaces.

g)   Because of the sealing effect of Na-clays, the water movement is restricted. The soil a few centimetres below the surface may therefore be almost saturated with water while the surface is dry and vice versa.

h)   Runoff water is always turbid and stays so for days together due to high dispersion of clay dominated by exchangeable sodium.

i)   Movement of clay from the surface to the lower layers results in the formation of hard impermeable clay pans.

j)   The soils contain free $CaCO_3$ in the surface layer and kankar pan at variable depths.

k)   Quality of underground waters is normally good but in some areas these waters may contain high CO_3^{2-} and HCO_3^- and their SAR and RSC may be high.

3.4. Measuring Sodicity Status

3.4.1. Exchangeable Sodium Percentage

The amount of sodium adsorbed on the soil, expressed in milliequivalent per 100 g soil, as percentage of soil CEC, is called exchangeable sodium percentage (ESP).

$$ESP = \frac{Exchangeable\ Na}{CEC} \times 100 \tag{1}$$

For this one needs to determine exchangeable Na and CEC of the soil. Cation exchange capacity is commonly determined using buffered salt solutions like NH_4OAc (pH 7.0) or NaOAc (pH 8.2). The CEC, which is a total of charges, i.e., variable charge due to amorphous oxides and organic matter and permanent charge due to clay minerals, is affected by the pH of the soil. For this reason, CEC of an alkali soil that is determined at lower pH than the pH of the soil is normally underestimated. Furthermore, the release of non-exchangeable Na from zeolites, the Na-bearing minerals present in most of the alkali soils, during extraction of exchangeable cations, complicates the computation of ESP by overestimating the exchangeable sodium. This results in a sum of exchangeable cations that exceeds the NH_4OAc-CEC. Moreover, when soil contains $CaCO_3$ or $CaSO_4$, determination of CEC vis-à-vis exchangeable cations becomes all the more difficult.

To take care of these difficulties, it is suggested that CEC of the salt-affected soils should be determined at the pH of the soil sample rather than at any arbitrary pH value. Though certain methods are available for determination of CEC of these problem soils, they are tedious and difficult to adopt for routine testing. Chhabra *et al.* (1975) using Ag-thiourea as index cation gave a method

Fig. 3.5.   Alluvial alkali soils have high pH throughout the soil profile and develop a pink colour when phenothphalene indicator is applied (Chhabra, unpublished).

Fig. 3.6.   Alkali soils have poor physical properties and develop cracks on drying (Chhabra, unpublished).

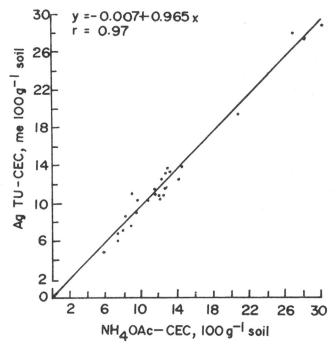

Fig. 3.7. Relationship between NH₄OAc and AgTU-CEC, me/100 g, of salt-affected soil (Chhabra *et al.*, 1975).

for determining CEC that is quick, accurate, and economical and gives reproducible results when compared to the NH₄OAc method (Fig. 3.7). However, this method takes into account pH-dependent CEC only up to pH 8 and thus cannot be used for alkali soils.

3.4.2. Sodium Adsorption Ratio

To overcome the above difficulties, efforts have been made to estimate ESP from the analysis of the saturation paste extract. Sodium adsorption ratio (SAR) of the saturation extract adequately defines the exchangeable sodium ratio (ESR) and ESP of the soil.

$$SAR = \frac{Na^+}{\sqrt{(Ca^{2+} + Mg^{2+}/2)}} \quad \text{(conc. } Na^+, Ca^{2+}, Mg^{2+} \text{ in me/l of saturation}$$

$$\text{extract)} \qquad (2)$$

$$ESR = \frac{Na^+}{Ca^{2+} + Mg^{2+}} \quad \text{(conc. } Na^+, Ca^{2+}, Mg^{2+} \text{ in me/100 g of soil)} \quad (3)$$

The empirical relationship between the two is given by Fig. 3.8 and expressed as:

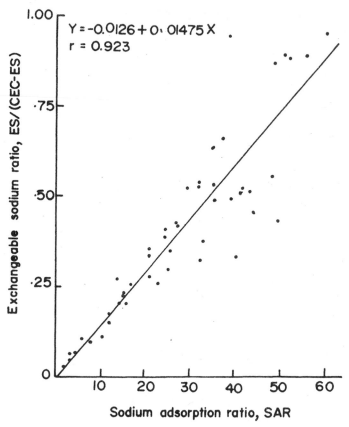

Fig. 3.8.   Relationship between SAR of the saturation extract and ESR of the soil (Richards, 1954).

$$ESR = -0.0126 + 0.01475\ SAR \qquad (4)$$

Now, as in alkali soils CEC is mainly due to Na^+, Ca^{2+}, and Mg^{2+}, equation 3 can be rewritten as:

$$ESR = \frac{ES}{Ca^{2+} + Mg^{2+}} \quad \text{(where ES = exchangeable } Na^+)$$

$$= \frac{ES}{CEC - ES}$$

$$ESR\ (CEC - ES) = ES$$

$$ESR \times CEC - ESR \times ES = ES$$

$$ESR \times CEC = ES + ESR \times ES$$

$$ESR \times CEC = ES\ (1 + ESR)$$

$$ES = \frac{ESR \times CEC}{1 + ESR}$$

$$\frac{ES}{CEC} = \frac{ESR}{1 + ESR}$$

$$\frac{ES}{CEC} \times 100 = 100 \times \frac{ESR}{1 + ESR}$$

$$ESP = 100 \times \frac{ESR}{1 + ESR} \tag{5}$$

$$ESP = \frac{100\ (-0.0126 + 0.01475\ SAR)}{1 + (-0.0126 + 0.01475\ SAR)} \tag{6}$$

Using this relationship the Soil Science Society of America has recommended a nomogram (Fig. 3.9) for estimating ESP. One should clearly see that the relationship is valid only for alkali soils and only up to a SAR of 30. Above this

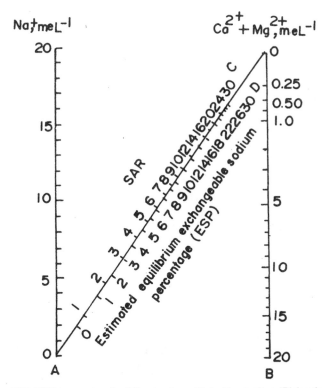

Fig. 3.9.   SAR-ESP nomogram for SP extracts and irrigation waters (Richards, 1954).

and in saline soils, because of ion pair formation, this relationship becomes invalid.

3.4.3. pH as a Measure of ESP

Alkali soils have pH greater than 8.2. Recent studies have shown that there is an intimate relationship between ESP and pH of the saturation paste. As the pH increases, ESP also increases (Fig. 3.10). Since the pH of the saturation paste can be easily determined in the laboratory, this property can be used as an approximate measure of ESP (Table 3.3), which is otherwise a cumbersome determination. The pH of a 1:2 of soil water suspension (pH2) is normally one unit lower than the pH of the saturation paste (pHs) (Fig. 3.11). Though pHs and ESP are well correlated over a large range of ESP, reliable interpretation

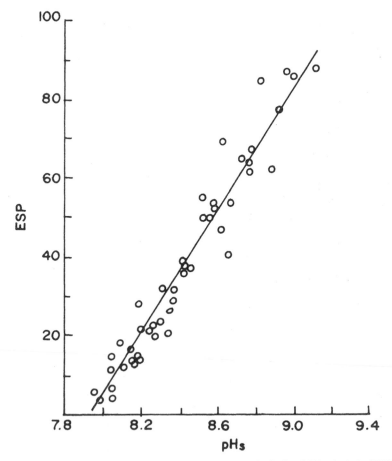

Fig. 3.10.   Relation between the pHs and ESP of an alluvial alkali soil (Abrol *et al.*, 1980).

Table 3.3. pHs-ESP relationship from a few studies

pHs	Exchangeable sodium percentage			
	Kanwar *et al.*, 1963	Kolarkar and Singh, 1970	Bhargava and Abrol, 1978	Abrol *et al.*, 1980
8.0–8.2	6–15	32–49	< 20	< 20
8.2–8.4	15–23	49–58	20–30	20–35
8.4–8.6	23–32	58–65	30–40	35–50
8.6–8.8	32–40	65–72	40–50	50–65
8.8–9.0	40–48	72–76	50–60	65–85
> 9.0	> 50	> 76	> 60	> 85

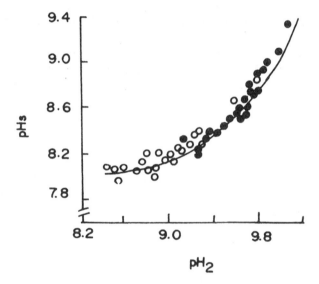

Fig. 3.11. Relation between the pHs and pH2 of an alluvial alkali soil (○, 0–15 cm; ●, 15–30 cm soil) (Abrol *et al.*, 1980).

of the data requires that the pH measurements are thoroughly standardized for any major group of soil.

3.4.4. Alkaline Earth Carbonates

Alkali soils contain soluble salts dominating in CO_3^{2-} and HCO_3^- of Na^+ and, almost invariably, concentration of Ca^{2+} is very low, about 2 me/l. Under natural conditions formation of alkali soils is possible only when most of the soluble and exchangeable Ca^{2+} precipitates in the form of $CaCO_3$ (calcite) and $CaMgCO_3$ (dolomite).

The presence of CO_3^{2-} and HCO_3^- can be easily detected using effervescence by dilute acid. The alkaline earth carbonates serve as a good source of native Ca for reclamation of alkali soils either through biological techniques or by acidulating amendments.

The fractional distribution of HCO_3^- and CO_3^{2-} species in a solution depends upon the dissociation constant of H_2CO_3 and pH as illustrated in Fig. 3.12.

$$H_2CO_3 \rightarrow HCO_3^- + H^+ \quad (pK1) \tag{7}$$

$$HCO_3^- \rightarrow CO_3^{2-} + H^+ \quad (pK2) \tag{8}$$

It is clear from the figure that at pH 8.2, nearly 98 per cent of the bulk fraction is represented by HCO_3^-, while at pH 10.5, more than 50 per cent of the bulk is represented by CO_3^{2-}.

If one knows the pH of the solution, then one can calculate the relative concentration of HCO_3^- and CO_3^{2-} species. For example, at pH 9, CO_3^{2-} /HCO_3^- ratio is 0.05, that means that CO_3^{2-} concentration is about 5 per cent of that of HCO_3^-. This is calculated as below:

$$pH = 9 \qquad H^+ = 10^{-9}$$

$$\frac{|\, CO_3^{2-} \,|\, |\, H^+ \,||}{|\, HCO_3^- \,|} = 5 \times 10^{-11} \text{ where } pK = 5 \times 10^{-11}$$

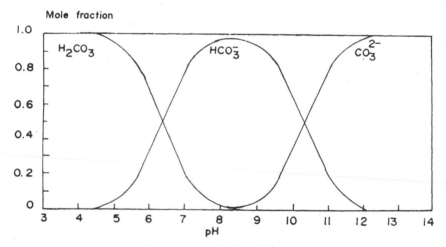

Fig. 3.12.   pH-dependence of carbonate species in solution.

$$\frac{|\ CO_3{}^{2-}\ |\ |\ 10^{-9}|}{|\ HCO_3{}^{-}\ |} = 5 \times 10^{-11}$$

$$CO_3{}^{2-} / HCO_3{}^{-} = 5 \times 10^{-11} \times 10^{-9} = 5 \times 10^{-2} = 0.05$$

3.4.5. Alkalinity of Alkali Soils

The alkaline reaction in alkali soils is due to the presence of $CO_3{}^{2-}$ and $HCO_3{}^{-}$. Their sum ($CO_3{}^{2-} + HCO_3{}^{-}$) is referred to as alkalinity of soils and is related to the carbon dioxide pressure in the gaseous phase. Under normal conditions, the soil water is in equilibrium with air (containing 0.35 mBar or 0.35×10^{-3} Bar carbon dioxide) and gives a critical pH of 8.5 corresponding to alkalinity concentration of about 1.5 to 2 me/l. However, when the same soil is water-logged and poorly aerated, which means that pCO_2 increases (due to its trapping in the soil) from 10^{-3} to 10^{-2}, the pH of the soil drops from 8.5 to 7.0 (Fig. 3.13). Such an increase in pCO_2 resulting in lowering of pH helps in increasing the solubility of calcite ($CaCO_3$) (Fig. 3.14). This gives us the possibilities to reclaim alkali soils by addition of organic matter, green manure,

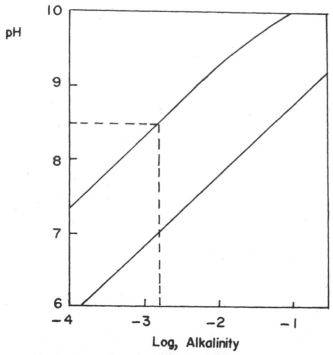

Fig. 3.13. pH dependence on alkalinity concentration ($CO_3{}^{2-}$ and $HCO_3{}^{-}$) at two CO_2 pressures: $10^{-3.5}$, well-aerated surface soil (top) and 10^{-2}, poorly-aerated subsoil (bottom).

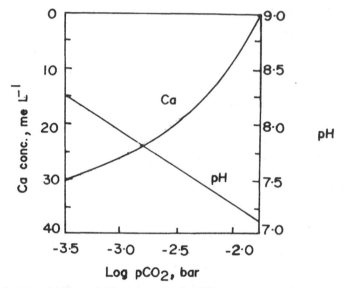

Fig. 3.14.   Calcite solubility and pH values at varying CO_2 pressures.

or even ponding water, as all these phenomena lead to higher CO_2 build-up and thus higher solubility of native Ca, which in turn is used to replace Na from the exchange complex.

3.4.6. Problem of Coloured Aqueous Extracts

Because of dissolution of a part of organic matter in Na_2CO_3, the aqueous extracts of alkali soils are often coloured and present difficulties in analysis of HCO_3^-, CO_3^{2-}, Cl^-, SO_4^{2-}, B, P and Ca^{2+} by conventional methods. This can be overcome by use of activated charcoal or precipitation of humic acid by lowering the pH. However, charcoal may create errors in the estimation of B and micronutrients especially if the pH of the extract is high. Lowering of pH is not possible if one is simultaneously determining CO_3^{2-} and HCO_3^- from the same extract. To avoid these complications, use of specific ion electrodes should be encouraged. For total analysis with respect to cations and P, the coloured solution can be dried and then digested in HNO_3-$HClO_4$ prior to processing.

3.5. Water Permeability and Concept of Threshold Electrolyte Concentration

When alkali soils are irrigated, silt and clay fractions rich in exchangeable Na^+ get dispersed and clog the pores, thereby reducing the air and water permeability of the soil (Fig. 3.15). This prevents proper gaseous exchange between soil and air, and leaching of Na^+ and other salts from the surface soil.

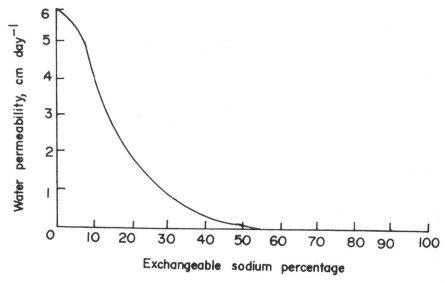

Fig. 3.15.   Effect of soil ESP on water permeability of a sandy loam alluvial soil.

For alkali soils, values of K, the saturated hydraulic conductivity, as low as 0.85 cm/day to nil have been reported from various field studies. The effect of ESP on the water permeability varies with the nature of clay minerals, being greater in swelling types and less in non-swelling types (Fig. 3.16).

The swelling of the minerals and their dispersion can be minimised by effectively controlling the thickness of the diffused double layer, by increasing the electrolyte concentration.

It is observed that if one leaches a soil of known ESP with a solution having a fixed SAR but increasing EC, one obtains a value of EC beyond which there is no further increase in permeability of the soil (Fig. 3.17). Hence, to maintain hydraulic conductivity at a particular ESP, one has to maintain a particular electrolyte concentration. This concentration of electrolyte below which there is a decrease in soil permeability at a given ESP is referred to as the threshold electrolyte concentration. In practice, normally one takes the electrolyte concentration at which one can maintain 75 per cent of the permeability.

If one plots this threshold electrolyte concentration against ESP of the soil, one gets the relationship expressed in Fig. 3.18. This figure defines two zones, one of stable and the other of decreasing permeability in terms of combination of ESP-threshold electrolyte concentration. In terms of SAR (Fig. 3.19) the relationship is expressed by the following equation:

Threshold conc. (me/l) = 0.60 + 0.56 SAR   (9)

The equation, based on Gapon's equation, does not contain any soil characteristics and thus can be regarded as a useful guideline. Basically, it can be used in two cases:

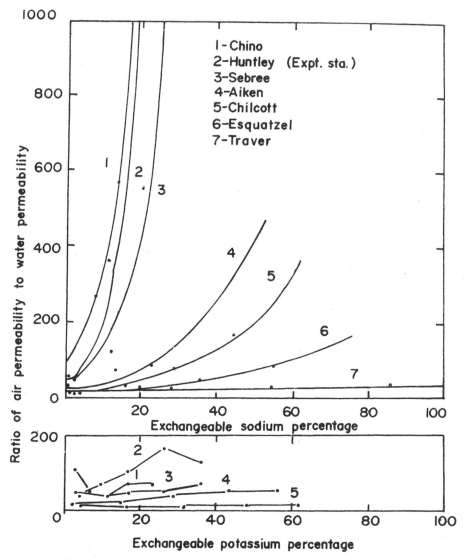

Fig. 3.16.   Effect of exchangeable sodium and potassium percentage on the air-to-water permeability ratio of soils: numbering (7-1) refers to increasing levels of swelling clays
(Richards, 1954).

a)   When the SAR value of SP extract of the soil is known, the electrolyte level
required in the irrigation water to maintain permeability can be predicted.

b)   When a non-alkali soil is being irrigated with irrigation water of un-
favourably high Na⁺ levels, this equation can be used to predict the
likely long-term effect of such irrigation water on the soil, i.e., when the

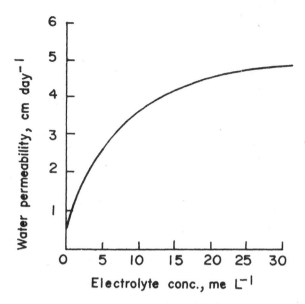

Fig. 3.17.   Effect of electrolyte concentration on the water permeability of an alkali soil (ESP 50).

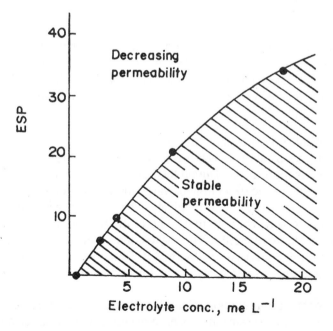

Fig. 3.18.   Threshold electrolyte concentration (me/l) of irrigation water as related to ESP of soils
(Rhoades, 1972).

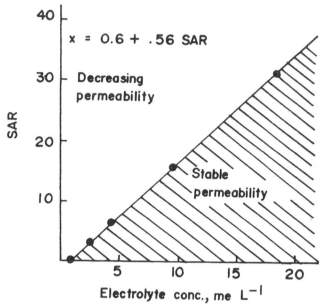

Fig. 3.19.   Basis for IW assessment using the concept of threshold electrolyte concentration and the Gapon equation (abscissa and ordinate can be characteristics of the IW) (Rhoades, 1982).

soil composition (ESP) will attain equilibrium with the irrigation water after long-term application. However, the time required to reach equilibrium is essentially dependent on three factors: (i) CEC of the soil, (ii) electrolyte concentration of the irrigation water, and (iii) rate of irrigation water application.

The practical implications of this are:

a)   When one is using high Na^+ waters during the off-rainy season, the soil becomes more sodic. During the rainy season, when the soil receives rain water, i.e., low electrolyte, water, then it immediately shows dispersion and reduction in permeability, which will not allow the salts to be leached down. Thus, to make use of the rain water to flush the salts, one has to increase the electrolyte concentration by applying gypsum on the surface. This will do two things: (i) maintain electrolyte concentration and thus permeability of the soil and (ii) reduce SAR and ESP of the soil.

b)   Another practical example relates to reclamation of saline soils. In the initial stages of soil reclamation, using drainage water alone, or mixing good quality irrigation water with more saline underground water, may accelerate the leaching process (due to higher permeability). As reclamation proceeds, the fraction of the saline water may then be reduced in successive steps. This will save a lot of good quality irrigation water.

3.6. Shear Stress and Erosion

When the soils become waterlogged and more so if accompanied by accumulation of salts on the surface, their shear stress decreases, resulting in more erosion. The critical shear stress, which is defined as the stress required to initiate erosion of soil, depends upon the type of clay, which in turn governs their swelling behaviour and CEC, moisture content, amount of organic matter, thixotrophy, and type and concentration of ions in the soil pores and eroding fluids. Arulanandan and Heinzen (1977) observed that critical shear stress was reduced from 25 to 5 dynes/cm^2 when SAR of the pore fluid increased from 5 to 15 for the same CEC of the soil. At still higher SAR of 60, the critical shear stress was further reduced to one dyne/cm^2 and became independent of CEC. The nature of cations and their proportion in the exchange complex affect the bonding forces and thus the cohesiveness of the soil. Increase in exchangeable sodium and pH of the soil results in greater dispersion, decreases the shear stress, and thus encourages the erosion. Relationship between moisture content, pH, and shear stress of an alluvial soil is described in Fig. 3.20 (Gupta, 1982). Thus, in alkali soils, the capacity of bunds, dykes, and other earthen structures to withstand erosion is very low. This decreases their life and causes extra expenditure on their proper maintenance.

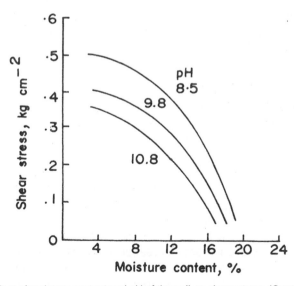

Fig. 3.20.   Effect of moisture content and pH of the soil on shear stress (Gupta, 1982).

4

Management of Alkali Soils

Management of alkali soils presents difficulties due to their physical and chemical properties, which affect seed-bed preparation, irrigation practices, drainage, choice of crops, and other field operations.

4.1. Factors Affecting Plant Growth in Alkali Soils

Plant growth, yield, and/or quality of crops in alkali soils may be affected by any of the following:

a) Poor water and air permeability as a result of high dispersion of soil aggregates and clay particles.

b) Low availability of water due to poor conductance from the lower soil layers.

c) Hard crust on the surface layer that greatly hinders seedling emergence and reduces the germination percentage and thus plant population.

d) Difficulties in preparing seed-bed as all the soil does not come to proper moisture conditions at the same time; this also results in the formation of clods and poor tilth of the soil.

e) Deficiency of Ca, since nearly all the soluble and exchangeable Ca is precipitated as insoluble $CaCO_3$.

f) Excess of Na, which is toxic *per se* to the plants and causes imbalance due to antagonistic effect on K and Ca nutrition.

g) Toxic concentration of HCO_3^- and CO_3^{2-} ions.

h) Decreased solubility and availability of micronutrients such as Zn and Fe, due to high pH, $CaCO_3$, and soluble HCO_3^- and CO_3^{2-} ions.

i) Increased solubility and accumulation of certain toxic elements such as F, Se, and Mo in plants that may affect crop yield and/or health of the animals feeding on them. Increased concentration of F can cause severe bone diseases in animals. Selenium toxicity is quite common among cattle fed on fodder or straw raised on alkali soils.

j)   Low activity of useful microbes due to high pH and excess exchangeable Na.

k)   Presence of a hard $CaCO_3$ layer, which is a physical barrier for the
movement of water and salts and the vertical root growth of crops and
trees.

l)   Continuous loss of fertile topsoil due to air and water erosion.

Under field conditions, plant growth is adversely affected by a combination
of the above factors, the extent depending upon the amount of exchangeable
sodium, pH, nature and stage of the crop growth, environmental conditions,
and overall management levels. Table 4.1 gives the approximate extent of
sodicity hazards in relation to ESP and pHs of the soil.

Table 4.1.   Exchangeable sodium percentage of soil and sodicity hazards to plants

Approximate ESP	pHs	Sodicity hazard
< 15	8.0–8.2	None to slight
15–35	8.2–8.4	Slight to moderate
35–50	8.4–8.6	Moderate to high
50–65	8.6–8.8	High to very high
> 65	> 8.8	Extremely high

Remarks: The adverse effect of exchangeable sodium on the growth and yield of crops in various
classes occurs according to their relative tolerance to sodicity. Whereas the growth and yield of
sensitive crops are affected at ESP levels below 15, only extremely tolerant native grasses grow
at ESP above 65. (Abrol *et al.*, 1980).

4.2. Crop Responses to Soil ESP

Crops vary widely in their tolerance to soil exchangeable sodium. In general,
rice and other cereals are more tolerant than legumes as they require less Ca,
availability of which is a limiting factor in alkali soils. Crops that can withstand
excess moisture conditions are generally more tolerant to alkali conditions.
Though the crop tolerance to soil sodicity varies with the stage of growth,
variety, environmental conditions, and level of management, their relative
tolerance is summarised in Table 4.2.

Among the cultivated crops rice is most tolerant (Fig. 4.1a) to soil sodicity.
It can withstand an ESP of 50 without any significant reduction in yield. It is
followed by sugarbeet and teosinte. Wheat (Fig. 4.1b), barley, and oat are
moderately tolerant. Legumes such as gram (Fig. 4.1c), mash, lentil (Fig. 4.1d),
cowpea (Fig. 4.2), chickpea, and pea are very sensitive and their yield
decreases significantly even when the soil ESP is less than 15. *Sesbania* is
an exception among the leguminous crops as it can grow at ESP up to 50

Table 4.2.   Relative tolerance of crops and grasses to exchangeable sodium

Tolerant	Moderately tolerant	Sensitive
Karnal grass (*Leptochloa fusca*)	Wheat (*Triticum aestivum*)	Gram (*Cicer arietinum*)
Rhodes grass (*Chloris gayana*)	Barley (*Hordecum vulgare*)	Mash (*Phaseolus mungo*)
Para grass (*Brachiaria mutica*)	Oat (*Avena sativa*)	Chickpea (*Cajanus cajan*)
Bermuda grass (*Cynodon dactylon*)	Shaftal (*Trifolium resupinatum*)	Lentil (*Lens esculenta*)
Rice (*Oryza sativa*)	Lucerne (*Medicago sativa*)	Soybean (*Glycine max*)
Dhaincha (*Sesbania aculeata*)	Turnip (*Brassica rapa*)	Groundnut (*Arachis hypogaea*)
Sugarbeet (*Beta vulgaris*)	Raya (*Brassica juncea*)	Sesamum (*Sesamum oriental*)
Teosinte (*Euchlaena maxicana*)	Sunflower (*Helianthus annuus*)	Mung (*Phaseolus aureus*)
	Safflower (*Carthamus tinctorius*)	Pea (*Pisum saccharatum*)
	Berseem (*Trifolium alexandrinum*)	Cowpea (*Vigna unguiculata*)
	Linseed (*Linum usitatissimum*)	Maize (*Zea mays*)
	Onion (*Allium cepa*)	Cotton (*Gossypium hirsutum*)
	Garlic (*Allium sativum*)	
	Pearl millet (*Penisetum typhoites*)	
	Cotton (*Gossypium hirsutum*)	

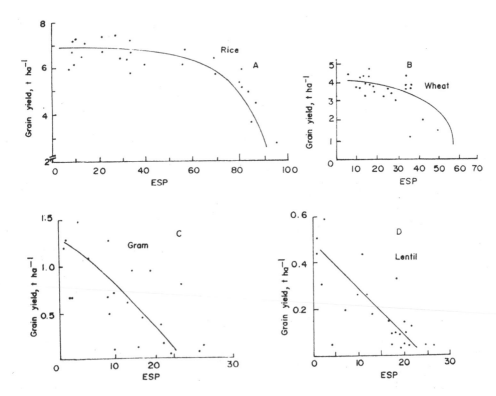

Fig. 4.1(A–D). Effect of soil ESP on the yield of crops (Abrol and Bhumbla, 1979).

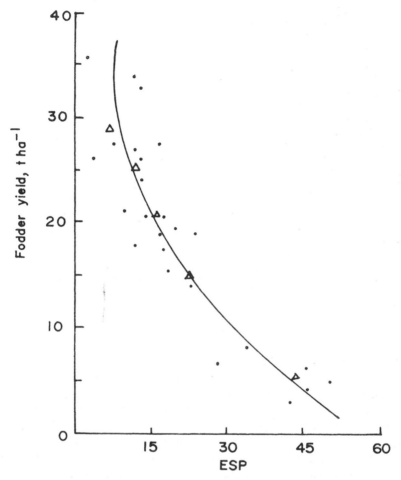

Fig. 4.2.   Effect of soil exchangeable sodium percentage on the fodder yield of cowpea (Singh *et al.*, 1980).

(Fig. 4.3) without any reduction in yield. It is, hence an excellent crop for green manuring in alkali soils.

Among the oilseed crops, except groundnut, all are moderately tolerant to soil ESP. However, raya followed by sunflower is more tolerant than others. The relative tolerance to soil ESP among the green fodder crops decreases in the following order: teosinte, oat, shaftal, lucerne, turnip, berseem.

Some of the natural grasses, such as Karnal grass and Rhodes grass, are very tolerant to soil sodicity, and, in fact, grow normally under high alkali conditions. Karnal grass likes both waterlogging and high ESP in its growing environment. As soon as the soil ESP decreases (as a result of reclamation),

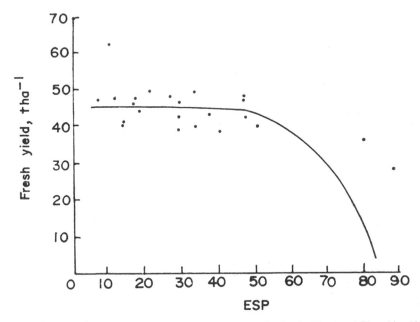

Fig. 4.3.   Effect of ESP on fresh biomass produced by *Sesbania* (Abrol and Bhumbla, 1979).

Karnal grass is not able to grow and disappears from the area. Grasses also ameliorate the soil and thus, after a few years of their cultivation, other field crops can be grown without the addition of amendments.

On the basis of earlier studies (Bernstein and Pearson, 1956; Pearson and Bernstein, 1958), Pearson (1960) listed crops according to their tolerance to exchangeable sodium (Table 4.3). A careful comparison of the listing in Table 4.2 with that in Table 4.3 brings out wide variations in the observed tolerance of some crops to exchangeable sodium. This is mainly due to the fact that:

a) The results of Pearson are based on soils whose structure had been stabilised with substances like krilium or other soil conditioners, a situation which is practically not possible under field conditions. Also, under field conditions crop growth in alkali soils is adversely affected by poor physical conditions as well as by direct toxic effect of Na and nutritional disorders brought about by high pH, $CaCO_3$, and high Na.

b) In Table 4.2 rice is rated as tolerant, while in Table 4.3 it is rated as moderately tolerant. This is mainly because Pearson used seeded rice, while work done in India refers to transplanted rice.

c) The results obtained in pots under controlled conditions are affected by only a uniform level of ESP, while under field conditions ESP of the lower layers, which is generally greater than that of the surface layers, also influences the crop performance. Thus, it is the combined effect of

Table 4.3. Tolerance of various crops to exchangeable sodium percentage under non-saline conditions

Tolerance to ESP and range at which affected	Crops	Growth response under field conditions
Extremely sensitive (ESP = 2–10)	Deciduous fruits Nuts Citrus (*Citrus* spp.) Avocado (*Persea americana* Mill.)	Sodium toxicity symptoms even at low ESP values
Sensitive (ESP = 10–20)	Beans (*Phaseolus vulgaris* L.)	Stunted growth at these ESP values even though the physical condition of the soil may be good.
Moderately tolerant (ESP = 20–40)	Clover (*Trifolium* spp.) Oats (*Avena sativa* L.) Tall fescue (*Festuca arundinacea* Schreb.) Rice (*Oryza sativa* L.) Dallisgrass (*Paspalum dilatum* Poir.)	Stunted growth due to both nutritional factors and adverse soil conditions.
Tolerant (ESP = 40–60)	Wheat (*Triticum aestivum* L.) Cotton (*Gossypium hirsutum* L.) Alfalfa (*Medicago sativa* L.) Barley (*Hordeum vulgare* L.) Tomatoes (*Lycopersicon esculentum* Mill.) Beet, garden (*Beta vulgaris* L.)	Stunted growth usually due to adverse physical conditions of soil.
Most tolerant (ESP > 60)	Crested and Fairway wheatgrass (*Agropyron* spp.) Tall wheatgrass (*Agropyron elongatum* Host Beau.) Rhodes grass (*Chloris gayana* Kunth)	Stunted growth usually due to adverse physical conditions of soil.

Source: Pearson (1960).

rootzone sodicity rather than a single ESP level that determines the relative tolerance of crops under field conditions.

4.3. Reclamation and Management of Alkali Soils

4.3.1. Amendments

To have successful crops in alkali soils, ESP of the soil must be lowered, which can be achieved by application of amendments. Amendments are those materials which (i) directly supply soluble Ca for the replacement of exchangeable Na^+ or (ii) furnish Ca indirectly by dissolving calcite, natively found in alkali soils, due to their acidulating action and/or increasing the partial pressure of CO_2. The Ca^{2+} so mobilised is used to replace Na^+ from the exchange complex and this reclaims the alkali soil.

$$2\,Na^+\text{-soil} + CaSO_4 \rightarrow Ca^{2+}\text{-soil} + Na_2SO_4\downarrow \tag{1}$$

Alkali soil   Gypsum   Normal soil

4.3.2. Chemical Amendments

Amendments like gypsum ($CaSO_4 \cdot 2H_2O$) and calcium chloride ($CaCl_2 \cdot 2H_2O$) directly supply Ca^{2+} to replace exchangeable Na^+ and thereby lower ESP. Mine gypsum is the most commonly used amendment for alkali soil reclamation.

Pyrites (FeS_2) furnish Ca^{2+} indirectly, by first oxidising to an acid which in turn reacts with soil lime to furnish soluble calcium.

$$2\,FeS_2 + 2\,HOH + 7\,O_2 \rightarrow 2\,FeSO_4 + 2\,H_2SO_4 \tag{2}$$

$$CaCO_3 + H_2SO_4 \rightarrow CaSO_4 + CO_2 + H_2O \tag{3}$$

$$2\,Na^+\text{-soil} + CaSO_4 \rightarrow Ca^{2+}\text{-soil} + Na_2SO_4\downarrow \tag{4}$$

The oxidation of pyrites starts as soon as the material is mined and exposed to air. The reaction even though slow is wasteful as some of the S may escape to atmosphere as follows :

$$FeS_2 + 3\,O_2 \rightarrow FeSO_4 + SO_2\uparrow \tag{5}$$

To minimise this wastage, it is important that its oxidation is encouraged under moist conditions where SO_2 can be directly converted into H_2SO_4 and used for reclamation of soil. This oxidation process is mediated through the bacteria *Thiobacillus ferrooxidens*, which requires a warm, well aerated and moist soil and low pH of 3 to 4 for efficient working. However, when pyrites is added to the alkali soil, the acid already contained in it reacts with $CaCO_3$ of the soil to release Ca^{2+} for exchange with Na^+ on the clay complex. Due to this and high alkalinity of the soils, pH in the vicinity of pyrites granules increases which decreases the biological oxidation of pyrites drastically. As a result of this, oxidation of S in pyrites is never complete and thus its efficiency remains much below that of gypsum when applied on equivalent S basis. Many studies (Table 4.4) have shown that its efficiency is about 70 to 80 per cent as compared to gypsum at the recommended dose of application. Since efficiency of pyrites as an amendment is governed largely by its water soluble S prior to its application in the soil, methods to increase oxidation of pyrites through incubation with FYM and good soil or even storing finely ground pyrites under moist conditions for a reasonably long period will give better results. For the same reason during transportation and storage of pyrites, it should be well

Table 4.4. Effect of gypsum and pyrites on the yield of crops in alkali soils

Source	Soil type and pH	Crops	Grain yield, t/ha, at different levels (% GR) of amendments									
			0		25		50		75		100	
			G	P	G	P	G	P	G	P	G	P
Abrol and Verma (1978)	Alfisol, initial pH 10.5	Rice	3.86	3.86	6.71	5.71	6.81	6.09	7.44	6.71	7.24	6.91
		Wheat	0.02	0.02	1.46	0.15	3.15	0.54	3.60	1.35	4.22	1.35
Chauhan (1978)	Alfisol, initial pH 9.2	Rice	1.73	1.73			2.77	0.54				
		Barley	2.62	2.62			3.67	3.64				
Bhatt et al. (1978)	Alfisol, Farmer's field, pH varying from 9.0 to 10.5	Rice	2.93	2.93	3.23	3.49	3.57	3.87	3.13	4.06	4.09	4.87
		Wheat	1.50	1.50	1.18	1.18	1.74	1.27	1.58	1.60	2.33	1.86
Singh et al. (1981)	Alfisol, initial pH 10.2	Rice	2.61	2.61		6.65[1]	6.87			7.59[2]		7.66
		Wheat	2.05	2.05		3.72	4.80			4.84		5.00
Sharma and Gupta (1986)	Vertisol, initial pH 8.5	Rice	0.14	0.14	0.97	0.84	1.77	1.60	2.50	2.18		
		Wheat	0.50	0.50	1.00	0.88	1.67	1.51	2.40	3.22		
Chhabra et al. (1989)	Alfisol, initial pH 10.2	Rice	1.75	1.75	2.93	2.09	3.06	2.37			3.43	2.60

G, Gypsum; P, Pyrites; [1] 33% GR; [2] 66% GR.

protected from rains otherwise soluble S will be washed away leading to lower efficiency of rainwashed material. After application in the field, instead of continuous ponding of water, giving cycles of alternate wetting and drying helps to increase its oxidation and thus efficiency.

As with pyrites, conversion of mineral sulphur to H_2SO_4, which also takes place through the biological oxidation by *Thiobacillus thiooxidens*, is very slow in alkali soils.

$$2 S + 3 O_2 + 2 H_2O \rightarrow 2 H_2SO_4 \qquad\qquad (6)$$

For this reason, the relative efficiency of mineral S is also not high in alkali soils.

Sulphuric acid reacts rapidly with soil lime, since it does not have to go through the oxidation process. However, it is highly corrosive and dangerous to handle.

Chemical amendments cost money. The benefit of any amendment should be tested in field trials as to cost, safety in use, and effectiveness. Theoretical amounts of various amendments to supply Ca equivalent to 1 t of gypsum is given in Table 4.5.

Table 4.5. Equivalent amount of various amendments for supplying Ca in terms of pure gypsum

Amendment	Tonne(s) equivalent to 1 t of 100% gypsum[*]
Gypsum ($CaSO_4 \cdot 2 H_2O$)	1.00
Calcium chloride ($CaCl_2 \cdot 2 H_2O$)	0.85
Calcium nitrate ($Ca(NO_3)_2 \cdot 2 H_2O$)	1.06
Pressmud (lime-sulphur, 9% Ca + 24% S)	0.78
Sulphuric acid (H_2SO_4)	0.61
Iron sulphate ($FeSO_4 \cdot 7 H_2O$)	1.62
Ferric sulphate ($Fe_2(SO_4)_3 \cdot 9 H_2O$)	1.09
Aluminium sulphate ($Al_2(SO_4)_3 \cdot 18 H_2O$)	1.29
Sulphur (S)[**]	0.19
Pyrites (FeS_2, 30% S)[**]	0.63

[*]The quantities are based on 100% pure materials. If it is not 100% pure, then necessary corrections must be made. Thus, if gypsum is only 70% (agricultural grade gypsum), then the equivalent quantity to be applied will be

$$\frac{100 \times 1}{70} = 1.43 \text{ t} \qquad\qquad (7)$$

[**]Oxidation of 100% S is assumed. In practice this does not happen and thus its effectiveness is much lower than that of other amendments.

The replacement of cation in soil is governed by the type of cation (its valency, size, and degree of hydration), ion concentration, nature of ions associated with the cation, and nature of colloidal particles. Since H^+ has the same valency as that of Na^+ but is smaller in size and non-hydrated, it has greater replacing power and is preferred over Na^+ in the exchange complex. Thus, H^+ of the H_2SO_4 formed by the oxidation of pyrites, or directly supplied by H_2SO_4 as such, may also directly replace Na^+ from the exchange complex and lower ESP. Sharma and Gupta (1986) observed that though ESP decrease was similar in H_2SO_4 and gypsum treatments (Table 4.6), the saturated hydraulic conductivity of the soil was less and the water-dispersible clay greater in H_2SO_4 treatments. Similarly, the ESP drop and improvement in physical properties was greater when $Al_2(SO_4)_3$ was applied as amendment. This seems due to Ca^{2+} and Al^{3+} of the gypsum and $Al_2(SO_4)_3$ respectively, which are' more effective flocculents than H^+ of H_2SO_4. Furthermore, $Al_2(SO_4)_3$ and $CaSO_4$ solubilise slowly and, as a result, a major part of theme is not immobilised by soluble $NaHCO_3$ and Na_2CO_3 but is effectively used for replacing exchangeable sodium.

Table 4.6.   Effect of different amendments (applied at a rate of 100% GR) on physicochemical properties of a sodic vertisol

Amendment	pH (1:2)	EC, dS/m	ESP	Saturated hydraulic conductivity, mm/hr	Water-dispersible clay, %	Dispersion ratio
Control	8.8	9.80	65	0.06	37.2	0.41
Gypsum	7.9	0.72	14	4.77	8.0	0.09
Pyrites	8.0	0.31	20	1.64	32.4	0.36
H_2SO_4	7.5	0.18	14	2.98	30.4	0.54
$Al_2(SO_4)_3$	7.6	0.27	8	4.49	8.6	0.09
$FeSO_4$	7.9	0.85	21	1.59	33.7	0.37

Source: Sharma and Gupta (1986).

4.3.3. Organic Amendments

Alkali soils are generally low in organic matter. Addition of crop residues and other organic materials in the soil is beneficial as these help to improve and maintain soil structure, supply needed plant nutrients, prevent soil erosion, and hasten the reclamation process. Most of the organic amendments such as straw, rice husk (rice hulls), groundnut and safflower hulls, farmyard manure, compost, poultry droppings, green manure, tree leaves, and sawdust on their decomposition produce high partial pressure of CO_2 and organic acids which help increase the electrolyte concentration and lower the soil pH. These processes increase the solubility of calcite and thereby lower the ESP (Put-taswamygowda and Pratt, 1973). To achieve maximum benefits from organic amendments, submerged conditions should be maintained by continuously ponding water during the course of their decomposition. Singh (1974) observed that when straw of barley and *Sesbania* was decomposed under aerobic conditions, their efficiency in terms of reduction in ESP and improvement in soil physical properties (Table 4.7) was less than when they were decomposed under saturated (anaerobic) conditions.

Table 4.7.   Effect of C:N ratio of added organic matter, at a rate of 2 per cent, and soil saturation on improvement in calcareous alkali silty clay soil

Treatment	pHs	ESP	Modulus of rupture, bars	Hydraulic conductivity, cm/hr	Na in leachate, me/l
Barley straw aerated	9.1	81.5	12.9	0.03	16.6
Sesbania straw aerated	8.8	68.5	11.8	0.04	26.3
Barley straw saturated	8.4	56.1	11.3	1.72	30.2
Sesbania straw saturated	8.3	47.2	7.5	1.29	43.0
Original soil	9.5	95.0			

Source: Singh (1974).

Because of their coarse texture and slow decomposition, these materials do not allow the pores to be clogged and physically keep the soil porous by maintaining channels and voids that improve water penetration. Thus, they also act as physical amendments.

Organic materials together with inorganic amendments are more cost-effective, hasten the reclamation process, and increase the yield; therefore, their use should be encouraged (Dargan *et al.*, 1976). When applied alone, relatively large quantities (20 to 400 t/ha representing 1/5 to 1/3 of the soil volume) have to be used for effective results. If the C:N ratio of these materials is very wide, as is the case with sawdust, rice husk, and barley straw, then they decompose slowly and may be less effective than *Sesbania*, which has a narrow C:N ratio. Under these conditions deficiency of N may also occur and need to be taken care of. Under certain situations salinity effects, especially with cowdung slurry, may become limiting factors for their continuous use.

4.3.4. Industrial Byproducts as Amendments

Industrial byproducts like phosphogypsum, pressmud, molasses, acid wash, and effluents from milk plants may be used to provide soluble Ca directly or indirectly by dissolving soil lime, for reclamation purposes. As these materials can be cheap and locally available, their use should be encouraged. However, care should be taken not to introduce toxic elements like F, which may be present in large amounts in products like phosphogypsum (Chhabra *et al.*, 1980c).

Amendments may be useful where soil permeability is less because of low salinity, i.e., when electrolyte concentration of the leaching water is less than the threshold electrolyte concentration, excess exchangeable Na^+, or CO_3^{2-}, HCO_3^- in the water. But these will not be helpful if the poor permeability is due to soil texture, soil compaction, restricted layers (hardpan, claypans), or high water table. If the crop is receiving adequate water for near maximum yield, addition of amendment will not increase yield but may make water management a little easier, though at an additional cost of amendment including its handling and application.

4.3.5. Quantity of Amendment

The amount of an amendment needed for alkali soil reclamation depends upon the amount of exchangeable sodium to be replaced, which in turn is governed by the amount of sodium adsorbed on to the soil (ESP and CEC), sodicity tolerance, and the rooting depth of the crop to be raised.

One milliequivalent (me) of Ca^{2+} is required to replace 1 me of Na^+ from the exchange complex. This will amount to 0.02 g Ca/100 g soil or 0.86 g of pure $CaSO_4 \bullet 2H_2O$ per kg soil. Taking 2×10^6 kg as weight of 0–15 cm soil layer, the amount will be 1.72 t of gypsum per ha per me of Na^+ to be replaced. Taking this as a base and considering the per cent purity and equivalent efficiency, the amount of various amendments required to achieve a desired decrease in ESP can be calculated. Normally, for the upper 15 cm of soil depth, the amount of gypsum required to achieve reclamation varies depending upon the soil ESP and texture, from 4 to 40 t/ha. For reclaiming the upper 30 cm of soil one may need about double that amount.

Gypsum requirement (GR) of a given alkali soil can be calculated from the data on CEC and ESP of the soil. For example, if the ESP of the soil is 70 and the CEC 25 me/100 g soil, then to reduce ESP to 15, one will require [(70–15)/100] $\times$ 25 = 13.75 me Ca^{2+}/100 g soil.

In the laboratory, gypsum requirement (GR) of the soil is determined by analysis as suggested by Schoonover (1952). This involves equilibrating a known weight of air-dry soil with a known large volume of saturated gypsum solution and determining, by back titration, me of Ca^{2+} adsorbed by the soil from the solution. From this one can calculate the amount of gypsum required for the given soil. However, the GR determined by this technique comes out to be very high, 30 to 60 t/ha, and is normally beyond the economic means of the farmers in many developing countries. Abrol *et al.* (1975) reported that soluble Na_2CO_3 of the soil may partly inactivate the added Ca and thus overestimate the GR in the Schoonover's method. They suggested that GR be determine on salt-free soil, i.e., after leaching the soil with 60 per cent alcohol, so as to get realistic estimates.

Yadav and Agarwal (1959) reported that a dose, that can reduce the ESP by 60 to 70 per cent is an economical level of gypsum. Extensive work done at the Central Soil Salinity Research Institute, Karnal, India, has shown that 10 to 15 t of gypsum containing 70 per cent $CaSO_4 \bullet 2H_2O$, which is about half of the actual GR determined by Schoonover's method, is sufficient to reclaim the surface 15 cm soil of 1 ha if one is to start with rice as the first crop

(Table 4.8). This is in view of the high tolerance of rice to soil ESP, its shallow root system, and its capacity to solubilise soil $CaCO_3$ to further reduce soil exchangeable sodium. Over time, ESP of the surface as well as deeper layers gets reduced so as to allow cultivation of other crops. However, to grow sensitive crops such as pulses one needs to add high amounts of amendments right in the initial years of reclamation.

Table 4.8. Yield (t/ha) of crops as affected by gypsum treatments in an alkali soil

Gypsum t/ha	Sesbania* 1970	Wheat 1970-71	Rice 1971	Wheat 1971-72	Rice 1972	Wheat 1972-73	Mash 1973	Lentil 1973-74	Pearl millet 1974	Gram 1974-75
0.0	0.44	0.00	4.39	1.52	7.18	1.35	0.052	0.28	1.06	0.124
7.5	3.99	1.89	6.21	2.79	7.23	1.96	0.249	0.82	1.33	0.468
15.0	4.35	3.49	6.39	3.45	7.13	2.43	0.318	1.16	1.96	0.688
22.5	4.70	4.16	7.08	3.72	7.17	2.26	0.381	1.36	1.81	0.904
30.0	4.50	3.79	6.66	3.98	7.03	2.74	0.419	1.40	1.94	0.928
LSD at P= 0.05	0.51	0.64	0.81	0.70	NS	0.50	0.123	0.35	0.52	0.440
Normal yield		4.50	7.00	4.50	7.00	4.50	1.00	2.00	3.00	2.000

*Green matter yield.
Initial soil pH = 10.3; ESP = 94 in surface 15 cm soil.
Source: Abrol and Bhumbla (1979).

Many farmers may even reclaim their alkali soils by simply ponding water and growing certain tolerant grasses and/or rice continuously. This is achieved through dissolution of native $CaCO_3$ by biological action of plant roots. Though slow and less efficient, it may be the only way for poor farmers, especially under situations when amendments are not easily available.

4.3.6. pH and the Gypsum Requirement

As discussed earlier, there exists a good relationship between pH of 1:2 soil water suspension (pH2), pH of the saturation paste (pHs), and ESP of the soil. It has been observed that for soils of the Indo-Gangetic plains such a relationship can form a sound basis for calculating approximate gypsum requirement of the soils varying in texture and thus CEC (Fig. 4.4). Similar relationships for different soils can be established for quick recommendations of their amendment needs.

4.3.7. Efficiency of Gypsum and Soluble Salts

All alkali soils contain measurable amounts of $NaHCO_3$ and Na_2CO_3 which under normal conditions may react with the added gypsum to neutralise and

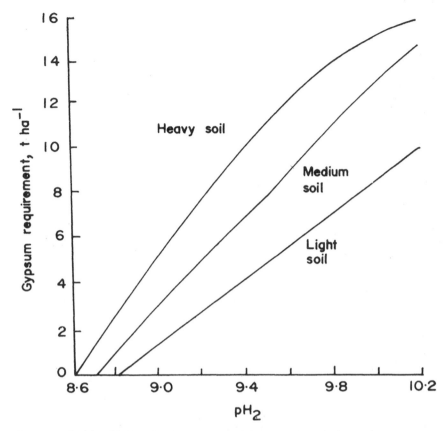

Fig. 4.4.   Relationship between pH$_2$ and gypsum requirement of alkali soils (Abrol *et al.*, 1980).

precipitate soluble Ca before it can be used to replace exchangeable Na$^+$, thereby decreasing the efficiency and increasing the amount of amendment needed to reclaim alkali soil.

$$2\,NaHCO_3 + CaSO_4 \rightarrow Ca(HCO_3)_2 + Na_2SO_4\downarrow \qquad (8)$$

If one leaches down the soluble salts prior to application of amendments, then most of the added soluble Ca can be used for replacing exchangeable Na$^+$ (Abrol and Dahiya, 1974) and thus large quantities of amendment can be saved. In view of this, under field conditions, gypsum should be applied after the first irrigation, which will ensure maximum leaching of soluble CO$_3^{2-}$ and HCO$_3^-$ and make them unavailable for precipitating soluble Ca of the added gypsum.

4.3.8. Method of Application

Amendment should normally be broadcasted and incorporated in the surface 10 cm soil. Field studies have shown that mixing a limited quantity of gypsum in deeper layers results in its dilution and therefore reduces soil improvement. Deeper mixing also results in higher amounts of gypsum being inactivated by soluble carbonates present in alkali soils, thereby further reducing the effectiveness of the applied gypsum and yield of crops (Table 4.9, Khosla et al., 1973).

Table 4.9.   Grain yield of crops as affected by quantity of gypsum and depth of mixing in an alkali soil

Gypsum level, t/ha	Mixing depth, cm	Grain yield, t/ha		
		Barley	Rice	Wheat
13.5	10	2.64	7.02	3.28
	20	2.46	6.05	3.16
	30	0.53	5.64	2.00
27.0	10	2.83	6.71	3.67
	20	3.22	6.28	3.67
	30	2.30	5.98	3.58

Source: Khosla *et al.* (1973).

To ensure proper dissolution of gypsum and leaching of replaced exchangeable Na^+, water is to be ponded on the soil. Normally, 10 to 80 per cent of the applied gypsum gets dissolved within 7 to 10 days of continuous ponding and leaching. However, when rice is to be taken as the first crop, which is normally recommended, post-application leaching is not needed and rice can be transplanted immediately after gypsum application. This is due to the fact that depth of water required to dissolve gypsum in soil and that too at high ESP is much less than that calculated on basis of its solubility in pure water. Further-more, water used for growing rice crop can itself be used for dissolving amendment and to leach down the reaction products from the soil. But when wheat, barley, or any other crop is taken as the first crop, sufficient irrigations, three to four of 7.5 cm each, should be given to dissolve gypsum, leach down the reaction products, and bring down the soil ESP to a desirable level before sowing.

When pyrites or S is used as an amendment, then a cycle of wetting and drying is needed to maintain proper moisture and aerobic conditions to ensure maximum oxidation and thus higher production of H_2SO_4.

4.3.9. Fineness of Gypsum

At the mine site gypsum is obtained in the form of lumps that need to be ground before application. Finer gypsum particles (Table 4.10) require less water for their dissolution (Hira *et al.*, 1981) and react faster. But very fine gypsum particles are prone to be wind-blown during application and to be inactivated by precipitation of $CaCO_3$ on their surfaces. Moreover, the cost increases with increase in fineness of the gypsum. Field and laboratory studies (Chawla and Abrol, 1982) have shown that gypsum lumps when ground to pass through a 2 mm sieve contain sufficient fine-sized particles to effectively reclaim alkali soil. A mixture of fine and medium-size particles maintains a steady electrolyte concentration, which is important for the maintenance of permeability of the low salinity alkali soils. Increasing the fineness of pyrites increases the specific surface area and results in better autooxidation of sulphur. Singh *et al.* (1981) observed an increase in water-soluble S content with increase in fineness of pyrites (Fig. 4.5).

Table 4.10: Effect of gypsum fineness on its dissolution in an alkali soil

Fineness of gypsum, mm	Amount of gypsum dissolved, g/100 ml	Depth of water involved, cm
< 0.10	2.00	2.50
< 0.26	1.43	3.50
< 0.50	1.05	4.75
0.26–0.50	0.59	8.50
0.50–2.00	0.42	12.00

Source: Hira *et al.* (1981).

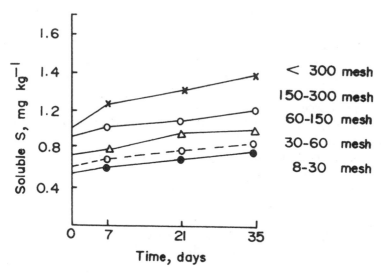

Fig. 4.5. Effect of particle size on the oxidation of pyrites (Singh *et al.*, 1981).

4.3.10. Frequency of Amendment Application

Alkali soils, once reclaimed, do not need repeated application of amendments provided they are continuously cropped. This is due to the fact that these soils contain high amounts of $CaCO_3$ that, together with Ca added from outside through fertilisers like calcium ammonium nitrate (CAN) and single superphosphate (SSP) and irrigation water, is sufficient to meet the Ca needs of the plants and to keep Na out of the soil exchange complex.

However, when large amounts of Na are added from outside through (a) flood waters from the adjoining unreclaimed alkali fields, (b) irrigation waters of high RSC, and/or (c) shallow brackish waters, the ESP of the soil will increase again. Similarly, under conditions conducive to precipitation of soil soluble Ca, such as prolonged dry fallow periods and irrigation with high RSC waters, the soil will become alkali again and may need repeated application of amendments.

4.3.11. Water Requirement for Gypsum Dissolution

The irrigation water requirements during reclamation of alkali soils have often been exaggerated on the premise that solubility of gypsum is low and it would need large quantities of water to dissolve the applied amendment. The solubility of gypsum in free water is 0.26 per cent at 25°C. On this basis USDA Salinity Laboratory Staff (Richards, 1954) recommended an application of 92 to 122 cm of water to dissolve 9 to 11.2 t/ha of agricultural grade gypsum of such fineness that 85 per cent of the material will pass through a 100 mesh sieve. Dutt *et al.* (1972) observed that solubility of gypsum was affected by a mixture of salts in the solution and the Na^+-Ca^{2+} exchange equilibria in the soil. Considering these points, they predicted that 52 to 72 cm of water was required to dissolve 16.5 to 24.0 t/ha of agricultural-grade gypsum. Oster and Halvorson (1978), however, reported that solubility of gypsum increased more than 10-fold when mixed in an alkali soil.

Gypsum solubility in alkali soils is affected by:
a) the activity of Ca^{2+} in solution, which in turn is governed by the solubility of solid $CaSO_4$ in the soil,
b) the size of gypsum particles, which affects its specific surface area and thus contact with water, and
c) the rate of Ca^{2+} diffusion from the solution to the exchange sites and rate of Na^+ diffusion away from the exchange sites.

$$CaSO_4 \cdot 2 H_2O \rightarrow Ca^{2+} + SO_4^{2-} + 2 H_2O \qquad (9)$$
$$\text{(solid phase)} \qquad\qquad \text{(solution phase)}$$

$$2 Na^+\text{--soil} + Ca^{2+} + SO_4^{2-} \rightarrow Ca^{2+}\text{--soil} + Na_2SO_4\downarrow$$

Soil exchange complex acts as a sink for Ca^{2+} released from the solid phase and, as Ca^{2+} released by gypsum dissolution is used up in replacing exchangeable Na^+, more of it is released from the solid to the solution phase. Thus, the amount of gypsum dissolved per unit of applied water is increased manifold. This process continues till the solution phase is saturated, i.e., the ion activity product of Ca^{2+} and SO_4^{2-} equals the solubility product of gypsum and this will happen only when all the Na^+ is replaced from the exchange complex. Abrol *et al.* (1979) and Hira *et al.* (1981) reported that the solubility of gypsum increased linearly with increase in ESP of the soil. A relationship of $Y = 0.0186\ X + 0.18$ ($r = 0.98$) between Y, i.e., mean per cent gypsum solubility, and X, i.e., ESP of the soil was observed by Abrol *et al.* (1979). Hira *et al.* (1981) reported that this relationship can be described by the equation $Y = 0.46\ X + 2.1$, where Y and X are the gypsum solubility and gypsum requirement of the soil, respectively.

On the basis of these concepts Hira and Singh (1980) observed that 4 cm of water was required to dissolve the added gypsum (< 0.26 mm size) and to reclaim an alkali soil (Fig. 4.6). They further reported that the water required to dissolve gypsum did not increase if the amount of gypsum added was equal to the gypsum requirement of the soil. At higher gypsum requirement, i.e., at high ESP, gypsum dissolution per unit amount of water increases significantly. However, to leach down the reaction products, a total of 26 cm of water, i.e., about four irrigations, is required in alkali soils.

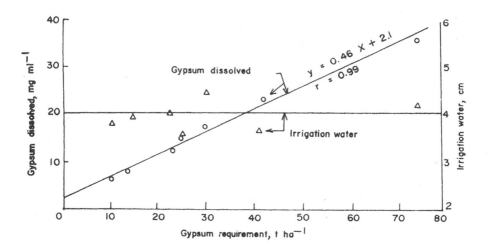

Fig. 4.6. Effect of gypsum requirement of the soil on the gypsum solubility and on irrigation water needs for gypsum dissolution (Hira and Singh, 1980).

4.5. Role of Rice as a Reclamative Crop

Growing of rice crop, along with other practices, has been observed to enhance the reclamation of alkali soils. The beneficial effect of rice can be attributed to the following reasons:

a) The rice plant can tolerate and, in fact, likes standing water during its growth (Fig. 4.7). This is a distinct advantage for crop growth in alkali soils since their poor air-water relationship adversely affects the growth of crops that cannot tolerate standing water and/or poor air-water permeability.

b) The water used for raising the rice crop is simultaneously used for dissolving the amendment and leaching salts.

c) Rice can tolerate a higher degree of exchangeable Na^+ and thus reclamation can be started even with a smaller amount of amendment.

d) Chhabra and Abrol (1977) from controlled greenhouse and laboratory studies observed a higher production of HCO_3^- in the leachate at all levels of ESP in the presence of rice plants than in their absence (Fig. 4.8). They reported that, because of submergence and biological activities of rice roots, partial pressure of CO_2 increases and this results in higher solubility of soil lime through its conversion into $Ca(HCO_3)_2$. Ca^{2+} of the solublised $Ca(HCO_3)_2$ lowers the ESP by replacing Na^+ from the exchange complex, which in turn is leached out as $NaHCO_3$ (Fig. 4.9) as per the following reaction:

$$CaCO_3 + CO_2 + H_2O \rightarrow Ca(HCO_3)_2 \hspace{2cm} (10)$$
(insoluble)   (biological)   (soluble)

$$2\ Na^+\text{–soil} + Ca(HCO_3)_2 \rightarrow Ca^{2+}\text{–soil} + 2\ NaHCO_3\downarrow$$

They further observed that, because of the activities of the rice roots, the pH of the soil as indicated by the pH of the leachate (Table 4.11)

Table 4.11.   The pH of the percolate as affected by presence or absence of growing rice plants in soils of varying ESP

Soil ESP	Treatment	pH of the percolate, on days after transplanting		
		15	35	65
93.3	Without plant	9.48	9.33	8.72
	With plant	9.43	9.30	7.97
46.0	Without plant	8.68	8.62	8.00
	With plant	8.63	8.37	7.68
29.9	Without plant	8.23	8.08	7.65
	With plant	8.18	8.13	762
10.5	Without plant	8.00	7.93	7.57
	With plant	7.82	7.82	7.42

Source: Chhabra and Abrol (1977).

Fig. 4.7.　Lowland transplanted rice is the most suitable crop for alkali soils (Chhabra, un-published).

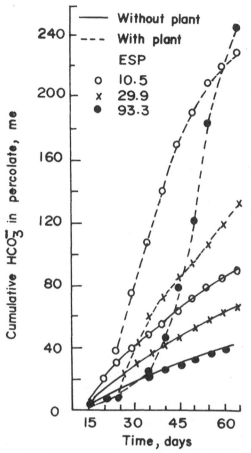

Fig. 4.8.   Effect of rice growth on the HCO_3^- content of the percolate from soils of different ESP (Chhabra and Abrol, 1977).

decreased sharply, which might have also increased the solubility of the native $CaCO_3$. Because of this decrease in pH, the dispersion of organic matter decreased and the leachate from treatments with rice became colourless. Thus, leaching of alkali soils in the presence of rice may also help in conserving the soil organic matter, losses of which otherwise will be quite substantial.

e) McNeal *et al.* (1966) observed that rice cultivation indirectly facilitates the removal of exchangeable Na^+ by increasing the cross-sectional area of conducting pores, resulting in increased permeability. Chhabra and Abrol (1977) observed that in highly alkali soils, rice roots provide channels for the movement of water, which increases permeability (Fig. 4.10) and helps in leaching of salts.

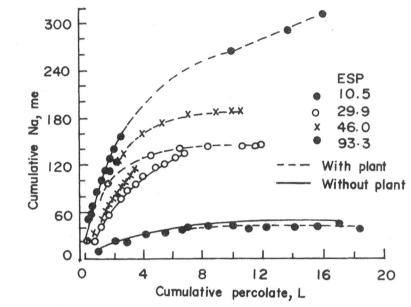

Fig. 4.9.   Cumulative soil Na removed as related to cumulated volume of percolate (Chhabra and Abrol, 1977).

f) Growing rice in poorly drained alkali soils helps in storing 12 to 15 cm of rain water right in the fields. This reduces the drainage needs of the area, minimises the peak flow rates, and helps recharge the groundwater (Narayana, 1979).

For these reasons, rice is not only an ideal crop for alkali soils but also enhances the reclamation process considerably. As a result of rice cultivation, ESP of not only the surface but also of the deeper layers decreases over a period of time (Fig. 4.11). It is observed, therefore, that during the first 4 to 6 years rice crop should be included in cropping sequence to be followed in alkali soils. Then, other crops sensitive to sodicity can also be grown profitably at a later stage.

4.6. Geohydrological Situations and Suitability of Soil for Rice Cultivation

Though rice is an ideal crop for poorly drained alkali soils, it should not be cultivated when a deep water table and high RSC water is the only source of irrigation, a situation that will lead to deterioration of surface soil properties and poor permeability. In such a situation, higher water use for lowland rice will lead to deterioration of soil physicochemical properties, which will be more

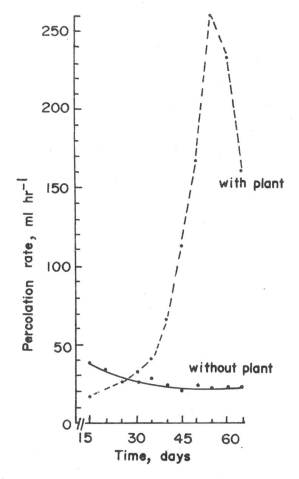

Fig. 4.10. Percolation rate as affected by rice growth in an alkali soil, ESP 93 (Chhabra and Abrol, 1977).

detrimental to the following upland crop. Geohydrological situations and the suitability of soil for rice cultivation are summarised in Table 4.12.

Rice-based cropping systems suitable for alkali soils are: rice-wheat, rice-Egyptian clover, and *Sesbania* (green manuring)-rice-wheat.

Rice-based cropping systems suitable for saline soils are: rice-rice, rice-barley, and rice-chilli.

Rice yield in the problem soils is affected by the method of its cultivation, (i.e., direct sowing through seeds or transplanting seedlings), nutritional disorders, adverse pH, drainage, and the genetic potential of the variety selected for the area. Some of the problems affecting rice yields are:

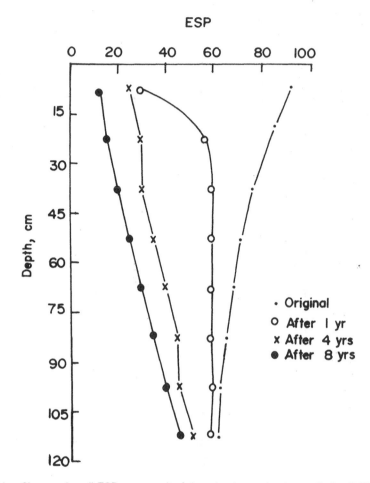

Fig. 4.11.   Changes in soil ESP as a result of rice-wheat cropping in an alkali soil (Chhabra, unpublished).

a)   Native low fertility or nutrient depletion or imbalances caused by intensive cultivation and injudicious fertiliser use.
b)   High salt contents leading to nutritional and osmotic stress and toxicities.
c)   Problems of low nutrient availability due to poor soil physical conditions and tissue injury associated with high pH.
d)   Soils with kankar or clay layers close to the surface obstructing root growth, salt removal, and drainage and causing nutritional problems.
e)   Low pH leading to nutritional problems and toxicities of iron and aluminium.

Table 4.12. Geohydrological situations and suitability of soil for rice cultivation

Geohydrological situation	Type of soil	Suitability for growing rice
Shallow water table of good quality, poor drainage	Alkali soil	Most suitable
Deep water table, good quality, plenty of canal water, poor drainage	Alkali soil	Suitable
Deep water table, high RSC, enough canal water	Alkali soil	Suitable
Shallow water table with hard pan close to surface, poor drainage	Alkali soil	Suitable
Shallow water table, brackish, no drainage	Saline soil	Suitable
Deep water table, brackish, not enough canal water	Saline soil	Unsuitable
Inundation due to tidal sea water, not enough fresh water but plentiful rains	Coastal saline soil	Suitable
Shallow water table with pyritic layer	Acid-sulphate soil	Suitable with special management only

f) Poorly drained high organic matter soils, with presence of toxic ions and oxygen stress.

g) Highly permeable soils having low water and fertiliser use efficiency.

h) Problems associated with topography and location, such as excessive water impounding or plant submergence, runoff, soil erosion, and poor drainage.

4.7. Cultural Practices

To ensure successful reclamation of alkali soils, the following practices should be adopted.

4.7.1. Land Levelling and Shaping

To ensure proper water management and uniform leaching of salts, the field should be levelled properly. To avoid major earthwork, the big fields should be divided into small parcels and levelled. Drastic removal of the surface soil will expose the subsoil containing $CaCO_3$, which can pose difficulties in reclamation and cropping of the area.

4.7.2. Plant Population

Because of the hard crust on the surface, germination percentage is often low in alkali soils. Plant population further decreases because of high rate of mortality, especially during early stages of plant growth. This together with poor tillering can reduce crop yield. Crop stand in alkali soils can be improved by increasing the seed rate and reducing the planting distance. In the case of rice, better plant population can be achieved by increasing the number of seedlings per hill and by gap-filling so as to replace the dead plants.

4.7.3. Age of Seedlings

Generally, crop tolerance increases with age. In rice, older seedlings (40–45 days) have been found to establish better than younger seedlings (25–30 days). Similarly, for establishing tree species, planting old seedlings has proved to be beneficial.

4.7.4. Green Manuring

Application of green manure can help to enhance OM content, increase partial pressure of CO_2, lower pH, enhance solubility of native $CaCO_3$, and add a considerable amount of plant nutrients in the soil. For this, *Sesbania*, which is tolerant to both high ESP and waterlogging, is an ideal crop. As it grows during the summer (May–June), a lean period for rice and wheat, it also fits into the cropping pattern to be followed in alkali soils.

Normally, 45-day-old *Sesbania* crop that attains a height of 1.5 to 1.8 m is ideal for incorporation as green manure crop. Being succulent and having a narrow C:N ratio (about 25), it decomposes very easily and quickly. The standing water in rice fields hastens the decay without any harm to rice crop. So there is no need to allow any time for the decomposition of the organic matter and rice seedlings can be transplanted immediately after the incorporation of *Sesbania* in the soil.

Since about 50 per cent of the organic nitrogen contained in the green manure crop is converted into the readily available ammonical form within 4 to 6 days and the rest within 10 to 20 days of incorporation (under the temperature conditions prevailing in the main rice-growing season), it helps ensure a steady supply of N to the rice crop.

A growth period of 45 days and "no decomposition period" results in better rice yields and a saving of 10 to 15 days for cultivation and decomposition of green-manuring crop. A longer growth period renders the plants hardy and difficult to decompose, whereas allowing more time for decomposition of a succulent crop like *Sesbania* in alkali soils results in greater losses of N through volatilisation and less N recovery by rice crop. It has been observed that yield increase in rice crop due to green manuring can be equivalent to about 80 kg of N applied through chemical fertilisers.

4.7.5. Continuous Cropping

On application of amendment, leaching (especially during growth of rice), and cropping, replaced Na^+ keeps moving downwards and there is continuous reduction in the exchangeable sodium of the soil throughout the soil profile.

Including Egyptian clover or other green fodder crop in the crop rotation, which requires frequent irrigations, one can further hasten this process. Therefore, the land should be continuously cropped to keep the downward movement of replaced Na^+ and soluble salts. Fallowing will encourage upward movement of salts and therefore reverse the process of reclamation.

4.8. Water Management

4.8.1. Drainage

As the alkali soils basically have low infiltration rate, all the rain water accumulates to create surface waterlogging. Even a low intensity shower or a normal irrigation may create temporary waterlogging and anaerobic conditions. Because of this, plants normally suffer oxygen stress in alkali soils. To avoid the problem, surface drainage, especially during the rainy season, is a must. During early stages of reclamation, surface runoff water that contains high concentration of soluble Na_2CO_3 and $NaHCO_3$ should not be allowed to pass on to the adjacent fields; it can cause sodication of the good quality land.

After the first few showers, the quality of the surface runoff water is generally very good. Instead of letting it go waste and contribute to the flood waters it may be stored in shallow ponds dug out at the lowest place at the farm. This stored water can be used for irrigation in the lean period.

If water table is high, which is the case in most soils, then sub-surface drainage has to be installed. However, because of low hydraulic conductivity, conventional sub-surface drainage to lower the water table is neither possible nor required in these soils. Normally, in alkali soils, the groundwater is of good quality and can be exploited for irrigation. For this, as many shallow cavity tubewells as possible should be installed to act as a source of irrigation and to provide vertical drainage. This has proved to be the most effective and economical way of controlling the water table and providing a source of irrigation.

Because of the poor hydraulic conductivity of these soils, natural recharge of the groundwater even during the rainy season is low. And as a result of vertical drainage through tubewells the water table goes down very fast in many areas. This sometimes even necessitates lowering of the motor for obtaining effective discharge and thus costs money to the farmers and exposes them to the poisonous gases that may accumulate in the pits (Chhabra, 1988c). To boost groundwater recharge, vertical bores with appropriate filters may be provided. This will also minimise runoff and conserve the rain water right in the area.

4.8.2. Irrigation

Irrigation management in alkali soils poses peculiar problems due to clogging action of dispersed soil particles and low stability of soil aggregates, which limit the water and air permeability of these soils. Because of this restricted entry from the surface, recharge of depleted soil moisture does not take place during irrigation of alkali soils. Furthermore, the amount of available water held by the soil as defined by the moisture content between field capacity (0.1 bar) and to wilting point (15 bars suction) is significantly reduced with increasing ESP (Abrol and Acharya, 1975). Thus, the available water storage capacity of the soil decreases with increase in exchangeable sodium.

Furthermore, during the post-irrigation period, the water transmission from the lower layers to the upper layers to meet the evapotranspiration needs of the crop is severely restricted by the low hydraulic conductivity of the soil. Acharya *et al.* (1979) observed that with increase in ESP of the surface soil, the water uptake by *Brassica juncea* was reduced drastically (Table 4.13). Not only that, the contribution of lower layers was low and practically nil when the ESP was high. Thus, at the same total water content, the available water is much less in alkali soils as compared to normal soils (Fig. 4.12).

Table 4.13. Water extraction by *Brassica juncea* from different depths as, related to the ESP of the surface soil

Water extraction	ESP of surface 15 cm soil				
	4	11	16	23	38
Extraction rate, cm day	0.23	0.24	0.20	0.18	0.14
Soil layer depth, cm	Per cent of total extraction				
0–15	46.4	40.0	67.5	75.0	78.0
15–30	30.0	31.1	28.0	25.0	22.4
30–45	16.3	18.7	4.5	0	0
45–60	7.3	6.2	0	0	0

*Mean of 17 days extraction period
Source: (Acharya *et al.* 1979).

Since only the upper soil layers are improved in the early stages of reclamation, the roots are confined primarily to the surface 30 cm layer. Thus, in addition to the limitations imposed by the physical properties of the soil, limited root growth restricts the depth of moisture extraction following irrigation. All these constraints affect the ability of alkali soil to supply adequate water to the growing plants.

As the depth of irrigation water and the interval between two irrigations depend upon the evaporative demand, nature of plant response to soil water stress, depth of root development, and the water storage and transmission characteristics of the soil, light but frequent irrigations should be applied in alkali soils. Heavy irrigation, which will supply water in excess of what can be absorbed by the soil in a few hours, can result in temporary waterlogging and oxygen stress in the rootzone, which can affect the crop growth and yield. This

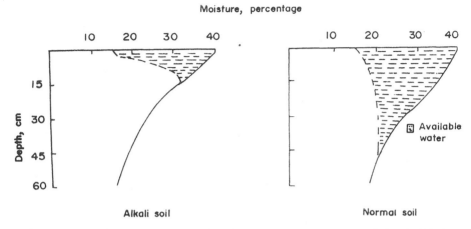

Fig. 4.12. Schematic diagram showing available water in a normal and an alkali soil (Chhabra, unpublished).

means that, the total amount of irrigation water remaining the same, it can be applied in small quantities but quite frequently. Generally, four to six irrigations of 5 cm will be required for a wheat crop. To facilitate this, big fields should be divided into small irrigation blocks.

Normally, a surface method of irrigation such as furrow or basin-type flood method is used for alkali soils. However, keeping in view their susceptibility to surface waterlogging, the sprinkler method could be promising because of its ability to supply water uniformly and in small quantities.

4.9 Distinguishing Features of Saline and Alkali Soils

The distinguishing features of saline and alkali soils are described below:

Characteristic	Saline soils	Alkali soils
A. Geographical distribution	Most often in arid and semiarid regions with rainfall less than 550 mm and in coastal areas.	Mostly in semi-arid and semi-humid regions with rainfall between 500 and 1000 mm.
B. Appearance	Generally with white salt efflorescence at the soil surface; patchy crop growth.	Soil surface covered with ash-coloured to brown or black salt efflorescence.
C. Natural vegetation	*Cressa cretica, Cyperus rotundus, Chloris pallida, Sporobolus pallida, Haloxylon salicornicum, Dichanthium annulantum, Suaeda fruticosa,*	*Sporobolus marginatus, Deostachya bipinnata, Suaeda maritma, Leptochloa fusca, Cynodon dactylon, Brachiaria mutica, Kochia indica,*

Characteristic	Saline soils	Alkali soils
	Butea monosperma, *Salicornia virginica,* *Tamrix* spp., *Atriplex* spp., *Phragmites communis,* Mangroves.	*Panicum antidotale,* *Prosopis juliflora,* *Butea monosperma.*
D. Chemical properties		
pHs	pH of the saturation paste is less than 8.2.	pH of the saturation paste is more than 8.2.
ESP	Exchangeable sodium percentage is less than 15.	Exchangeable sodium percentage is more than 15. In vertisols, even soils with ESP 6 may have alkali problem.
ECe	Saline soils have high amounts of soluble salts, which measured in terms of electrical conductivity of the saturated soils extract are always more than 4 dS/m at 25˚C.	Conductivity of the saturated soil extract is generally less than 4 dS/m at 25˚C. When soil has appreciable amounts of Na_2CO_3 and $NaHCO_3$, the surface soil ECe may be high.
Nature of soluble salts	Saline soils contain neutral soluble salts, mostly Cl^- and SO_4^{2-} of Na^+, Ca^{2+}, and Mg^{2+}. CO_3^{2-} is absent.	Alkali soils contain soluble salts capable of alkaline hydrolysis, mostly CO_3^{2-} and HCO_3^- of Na^+. Cl^- and SO_4^{2-} may also be present.
Nature of sparingly soluble salts	Soils may contain significant amounts of sparingly soluble gypsum.	Gypsum is nearly always absent. But all alkali soils contain free $CaCO_3$, the amount varying from traces to even more than 40 per cent.
pHs, ESP/SAR relationship	In general, there is no well defined relationship between pHs and ESP of the soil and SAR of the saturation extract.	For a given group of soils pHs increases with increase in ESP of the soil or SAR of the saturation extract; pHs is generally taken for approximation of ESP and sodicity status of the soil.
E. Physical properties		
Dispersion	Due to excess neutral soluble salts, the clay fraction is flocculated.	Excess exchangeable sodium and high pH result in high dispersion of clay fraction of alkali soils.
Permeability	Permeability of soils to air and water is generally good	Permeability of soils to air and water is restricted and becomes worse with increasing levels of ESP/pHs.
Water stagnation	Rain water does not stagnate if the water table is deep.	Rain water stagnates for long periods and even causes floods.
Colour of standing water	Generally clear.	Due to dispersion of clay particles, standing water is generally muddy.

Characteristic	Saline soils	Alkali soils
Effect of drying	On drying, the soils do not become too hard.	On drying, alkali soils become too hard to plough, develop craks, and often have thin crust on the surface
Plasticity	Normal.	Alkali soils get sticky and slippery when wet.
Erodability	Low to medium.	Alkali soils, because of their high degree of dispersion and low shear stress, are highly prone to erosion by both wind and water.
Presence of hardpan	Generally absent.	Alkali soils contain impervious $CaCO_3$ or clay pan in the soil profile, which generally acts as a barrier for the movement of salts and water, and vertical growth of roots.
F. Water table	Often shallow and is the main cause of waterlogging and salinisation.	Often moderate to deep; in areas with impermeable layers, perched water table may be observed.
G. Quality of groundwater	Generally poor; regions dominated by saline soils have mostly groundwater with high EC, which is a potential salinity hazard.	Groundwater in alkali soil regions is mostly low to medium in salts. In some cases it may have high RSC as a potential sodicity hazard.
H. Main adverse effects on plant growth	a) High osmotic pressure of soil solution leading to low water availability or physiological drought.	a) High pHs/ESP leading to poor permeability of air and water causing oxygen stress and low water availability.
	b) Direct toxicity of specific ions such as Cl^-, SO_4^{2-}, and B^-.	b) Direct toxicity of specific ions such as Na^+, CO_3^{2-}, HCO_3^{2-} F^-, and Se.
	c) Nutritional disorders.	c) High pHs causing induced nutritional deficiencies of Ca, Zn, Fe, and Mn.
	d) Oxygen stress.	d) Nutritional imbalances due to high Na^+, CO_3^{2-}, and HCO_3^-.
		e) Restricted vertical growth of roots due to presence of impervious layers.
I. Methods of reclamation	Saline soils are reclaimed essentially by removing excess soluble salts from the rootzone through leaching and drainage. Application of amendments may generally not be required.	Reclamation of alkali soils essentially requires lowering of pHs and replacement of exchangeable sodium by Ca through the use of amendments, and leaching of salts resulting from reaction of amendment with exchangeable sodium.

5

Principles Governing Fertilisation of Salt-affected Soils

Apart from reclamation by addition of amendments, leaching, and drainage, management of nutrients is of great importance for obtaining of high returns from salt-affected soils. Most often these soils have poor fertility status. In addition to providing the essential nutrients, harmful effects of excess salts or specific ions can be reduced, to some extent, by increasing the fertility of soil and thus help the plant survive the stress conditions.

Principles governing the application of fertilisers, as also the ameliorating techniques vary greatly with the nature of the soil problem.

5.1. Alkali Soils

High pH, excess exchangeable sodium, substantial presence of $CaCO_3$, soluble $NaHCO_3$, and poor soil physical conditions affect the solubility, chemical forms, transformation, and availability of many essential plant nutrients in alkali soils. Therefore, almost all the crops grown in such soils respond not only to the application of different fertiliser nutrients but also to the methods of their application. Out of 16 elements required for plant growth, alkali soils are generally deficient in Ca, N, Zn, Fe, and Mn.

5.1.1. Calcium

Alkali soils are deficient in both soluble and exchangeable calcium. Though these soils contain $CaCO_3$ ranging from traces to 40 per cent and more, the availability of Ca from it is insufficient to meet the plant's needs because of its low solubility at high pH (Table 5.1).

Moreover, because of the antagonistic effect of Na on Ca, increasing soil sodicity nearly always results in decreased uptake of Ca by plants (Fig. 5.1, Table 5.2).

Table 5.1.   Solubility of CaCO₃ as affected by the pH of water

pH	Solubility of CaCO₃, me/l
6.21	19.3
6.50	14.4
7.12	7.1
7.85	2.7
8.60	1.1
9.20	0.8
10.12	0.4

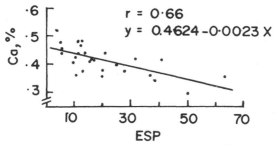

Fig. 5.1.   Calcium content of sunflower grain as affected by ESP of the soil (Chhabra *et al.*, 1979).

Table 5.2.   Effect of soil ESP on per cent Na and Ca contents of some 30-day-old plants

ESP	Sunflower		Safflower		Linseed		Cowpea		Raya	
	Na	Ca	Na	Ca	Na	Ca	Na	Ca	Na	Ca
7.6	0.09	2.78	1.01	1.36	1.48	0.46	0.16	2.35	0.50	2.98
12.5	0.10	2.87	1.42	1.22	1.53	0.46	0.24	2.33	0.73	2.91
16.6	0.26	2.68	1.85	1.28	1.76	0.44	0.25	2.24	1.00	2.80
23.0	0.41	2.62	2.28	0.88	2.10	0.34	0.32	0.05	1.31	2.35
44.2	0.52	2.25	2.81	0.63	2.40	0.27	0.66	1.72	3.02	1.84

Source: Chhabra *et al.* (1979); Singh *et al.* (1979, 1980, 1981).

Therefore, plants grown in alkali soils suffer more often from lack of calcium than from the toxic effects of excess Na (Chhabra and Abrol, 1983). However, in soils of moderate sodicities and in Na-sensitive crops, accumulation of Na in plant tissues in toxic quantities is chiefly responsible for poor growth and poor yields. Physiological disorders such as bad opening of cotton balls and fruit-end rot in tomato are ascribed to Ca deficiency. Absolute content of Ca (Fig. 5.2) and Ca:Na ratio (Fig. 5.3) are important indices for monitoring healthy growth of plants and ultimate yield of most crops grown in alkali soils (Chhabra and Abrol, 1975; Chhabra *et al.*, 1979; Singh *et al.*, 1979). Mehrotra and Das

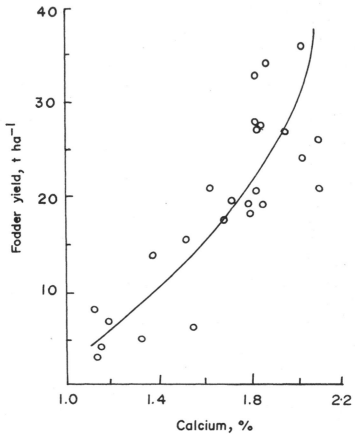

Fig. 5.2. Fodder yield of cowpea as affected by Ca concentration of plant grown in an alkali soil (Singh *et al.*, 1980).

(1973) reported that crops that have a narrower Ca:Na ratio under normal soil conditions are relatively more tolerant to alkalinity than those that have a broader ratio.

Gypsum, phosphogypsum, pressmud, and other amendments act as direct sources of Ca for plant growth. The availability of Ca from applied $CaSO_4$ depends upon the soil ESP, root CEC, and plant species. Poonia and Bhumbla (1972) using [45]Ca reported that 31 per cent of Ca in *Sesbania aculeata* (high root CEC) was contributed by the added $CaSO_4$, while in *Zea mays* (low root CEC) only 17 per cent was absorbed from it.

The acids and acid-forming amendments such as H_2SO_4, FeS_2, S, and $Al_2(SO_4)_3$ supply Ca by solubilising native $CaCO_3$. The effectiveness of such amendments depends upon their $CaCO_3$ solubilising capacity, which in turn

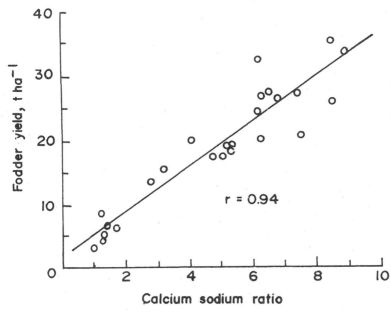

Fig. 5.3.   Relationship between Ca:Na and the yield of cowpea grown in alkali soils (Singh *et al.*, 1980).

is governed by the degree of acidity produced. Decomposing organic matter under anaerobic conditions helps in increasing pCO_2 and production of acids and acidic products, which increase the solubility of native $CaCO_3$.

Growing of crops and grasses increases the solubility of soil $CaCO_3$ and thus supplies Ca from native source. Chhabra and Abrol (1977) reported that the observed reclaiming effect of rice cultivation in alkali soils is mainly through the dissolution of $CaCO_3$ by the biological action of plant roots. While soil improvement can initially be achieved by application of amendments, growing of lowland rice crop can further improve the soil by gradually solubilising soil calcium carbonate, which in turn decreases the exchangeable sodium. Thus, use of adequate amounts of amendment and adoption of agricultural practices that increase the solubility of native $CaCO_3$ are essential for satisfactory crop yields and reclamation of alkali soils.

5.1.2. Nitrogen

Alkali soils are deficient in total and available nitrogen because of:

a) Low organic matter content

Since most alkali soils remain devoid of any vegetation for most of the year, addition of organic matter from natural vegetation in these soils is very low.

Most of these soils in tropical countries contain less than 0.1 per cent OM. As OM is the bank of nitrogen in the soil, alkali soils are poor in native available nitrogen (Chawla, 1969; Paliwal, 1972; Singh *et al.*, 1983).

b) Slow transformation of nitrogen

The biological transformation of nitrogen from amide to ammonia and nitrate is adversely affected by alkalinity and sodicity. Martin *et al.* (1942) reported that threshold pH values for nitrification of ammonia was 7.1 ± 0.1. They observed that nitrification did occur at higher pH values but was accompanied by considerable accumulation of nitrites (Fig. 5.4). Nitant and Bhumbla (1974) reported that complete hydrolysis of urea, the most commonly used nitrogenous fertiliser, in a soil of pH 9.8 was 4 days slower than that in soil of pH 8.6 (Fig. 5.5). This is attributed to the possible adverse effect of high pH on the activity of enzyme urease and/or the direct toxic effect of carbonate ions on the N-transforming bacteria. These factors delay and adversely affect the availability of applied N to the plants.

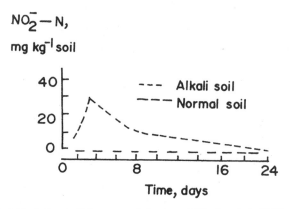

Fig. 5.4.   Effect of alkalinity on nitrite accumulation in soil (Martin *et al.*, 1942).

c) Volatilisation losses of added nitrogen

High pH/alkalinity and high amounts of $CaCO_3$ favour considerable volatilisation losses of applied N in alkali soils. Jewitt (1942) observed a loss of 87 per cent N from ammonium sulphate applied to a Barber soil (pH 10.5) in northern Sudan. Similarly, Bhardwaj and Abrol (1978) observed that nearly 32 to 52 per cent of the applied nitrogen was lost through volatilisation in alkali soils (Table 5.3). Rao and Batra (1983), from controlled laboratory studies, observed that 62 per cent of applied N to an alkali soil (pH 10.2) was lost against only 0.2 to 0.3 per cent from a normal soil (pH 7.8) through volatilisation in 10 days. These losses were, however, reduced to 32 per cent under waterlogged conditions.

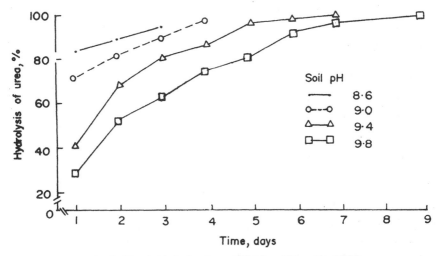

Fig. 5.5.   Effect of soil pH on hydrolysis of urea (Nitant and Bhumbla, 1974).

Table 5.3.   Effect of pH and CaCO₃ on the volatilisation losses of ammonical nitrogen from alkali soils

Soil pH (1:2 soil water suspension)	CaCO₃, %	Loss of applied N, %
9.0	nil	32.2
10.3	3.2	52.1
9.0	4.0	42.3

Source: Bhardwaj and Abrol (1978).

Folkman and Wachs (1973), Fleisher and Hagin (1981), and Rao and Batra (1983) have reported that loss of nitrogen as NH_3 through volatilisation is a function of total nitrogen applied and, therefore, is a first-order reaction. Thus, heavy applications result in greater losses of N and the availability of applied fertiliser N to the crop is reduced by 60 to 40 per cent.

Obrejanu and Sandu (1971) observed that, mainly because of these losses, the per cent response to applied N (Table 5.4) to sugarbeet crop at low level of gypsum application (high ESP/pH) was less than that at high level of gypsum application (low ESP/pH).

A considerable amount of N is lost through denitrification in alkali soils. Poor physical conditions of these soils favour frequent cycles of alternate wetting and drying, which promote N losses through denitrification.

d) Reduced symbiotic N fixation

In alkali soils, symbiotic fixation of atmospheric N is low. This is due to the sensitivity of the microbes to high pH/alkalinity (Mulder and Veen, 1960; Singh

Table 5.4.   Sugarbeet yield, t/ha, as affected by nitrogen at different levels of gypsum in a solonetic soil

Gypsum level, t/ha	Nitrogen applied, kg/ha			
	0	60	120	180
0	21(35)*	31(48)	38(50)	52(59)
8	27(45)	35(54)	51(66)	56(64)
16	60(100)	65(100)	77(100)	88(100)

* Figures in parenthesis are yields relative to highest gypsum level for a given nitrogen treatment
Source: Obrejanu and Sandu (1971).

et al., 1973) and reduced growth of most legumes, the host plants. Bhardwaj (1975) reported that though Rhizobia could survive in alkali soils of pH as high as 10.0, the effective contribution of the bacteria to the nitrogen needs of the plants was limited because of delayed nodulation and the sensitivity of the host plants to soil sodicity. Except for Sesbania aculeata, legumes are very sensitive to soil sodicity (Abrol and Bhumbla, 1979).

Because of low available N status and slow and wasteful transformation of organic and inorganic-N resulting in poor efficiency of added N, crops grown in alkali soils generally respond to higher levels of N (Fig. 5.6) than those raised in non-alkali but otherwise similar soil and climatic conditions (Kanwar et al., 1965; Bains and Fireman, 1968; Obrejanu and Sandu, 1971; Dargan and Gaul, 1974). In general, it is recommended that crops grown in alkali soils be fertilised at 25 per cent more N over the rates recommended for normal soils.

Studies by Nitant and Dargan (1974) showed that using $(NH_4)_2SO_4$ as a N source in alkali soils resulted in higher yields of rice and wheat, as compared to urea and calcium ammonium nitrate (CAN). This was attributed to the beneficial effect of residual acidity of this fertiliser. However, Rao and Batra (1983) reported that anions associated with ammonia do not influence the N losses. Field trials with rice-wheat rotation over a period of two years (Singh et al., 1983) did not show any significant differences in yield between $(NH_4)_2SO_4$ and urea when adequate quantities of amendments had been applied.

Efficiency of applied N can be increased by split application (Dargan et al., 1973; Singh et al., 1983), which decreases the losses of N through volatilisation. It has been observed in a number of field experiments that 33 per cent of total nitrogen applied as basal and the rest applied in two equal parts after 21 and 45 days of transplanting or sowing gives best results for rice and wheat grown in alkali soils. Foliar application of N (3 per cent solution of urea) together with a basal application (Table 5.5) gives good results and can save 40 to 60 kg N/ha in alkali soils (Dargan et al., 1973).

To avoid leaf burn, foliar application of urea should be done on a cloudy day and preferably in the afternoon.

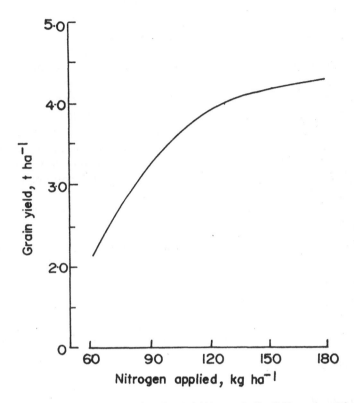

Fig. 5.6. Effect of nitrogen levels on the wheat yield in an alkali soil (Dargan and Gaul, 1974).

Table 5.5. Effect of soil and foliar application of urea on the yield of paddy and wheat in a partly reclaimed alkali soil

Treatment, kg N/ha	Mode of application	Grain yield, t/ha	
		Paddy	Wheat
Control[*]		4.36	1.95
20	Soil (1)[**]	5.21	2.51
40	Soil (2)	5.52	3.65
60	Soil (2)	5.58	4.53
20	Foliar (2)[***]	6.36	4.24
40	Foliar (4)	6.69	4.52

* Basal dose of 80 kg N through urea was applied to all the treatments.
** No. of splits.
*** No. of sprayings.
Source: Dargan *et al.* (1973).

Green manuring with "dhaincha" (*Sesbania aculeata*) has been observed to increase the grain yield of rice (Table 5.6) by an equivalent of 80 kg N/ha applied through urea. The effect of green manuring in improving crop yields is mediated through intensification of the reclamation process by increasing CO_2 evolution, improved physical conditions, mobilisation of Fe and Mn, and steady supply of nitrogen to the plants. Thus, green manuring should form an important component of any reclamation programme. For better results, crop grown for green manuring should be ploughed in after 45 to 60 days of growth and allowed to decompose under anaerobic conditions. This will help in higher production of organic acids and trapping of decomposition products for dissolution of soil $CaCO_3$. Rice seedlings can be transplanted, without any harmful effect, even just after the incorporation of green manured crop.

Table 5.6. Rice yield, t/ha, as affected by levels of N and green manuring with *Sesbania* in an alkali soil

Treatment	Nitrogen levels, kg/ha				Mean
	0	40	80	120	
Fallow (F)	2.64	3.86	5.56	6.01	4.52
Green manuring (GM)	5.64	6.31	7.30	7.71	6.74
LSD at P = 0.05,		N levels = 0.55,			F × GM = 0.39

Source: Dargan *et al.* (1975).

5.1.3. Phosphorus

Alkali soils normally have high extractable phosphorus status. Singh and Nijhawan (1943) reported high amounts of chemically extractable phosphorus in Bara soils of Punjab, India. Chhabra *et al.* (1980b) analysed a large number of soil samples from barren alkali soils of the Indo-Gangetic alluvial plains and reported that these soils contain high amounts of Olsen's extractable P, up to 134 kg P/ha, and that there is a positive correlation between the electrical conductivity and the extractable P status of the soil (Fig. 5.7). High amounts of Na_2CO_3 and $NaHCO_3$, which make up the bulk of soluble salts in these soils, react with native insoluble calcium phosphates to form soluble sodium phosphates and hence give a positive correlation between the electrical conductivity and their soluble P status.

$$Ca_3(PO_4)_2 + 3 Na_2CO_3 \rightarrow 3 CaCO_3 + 2 Na_3PO_4$$
(insoluble)   (soluble)

Furthermore, at high ESP, above 20, and at low level of fertiliser P application, there is a negative sorption of P by the alkali soils (Fig. 5.8). Thus,

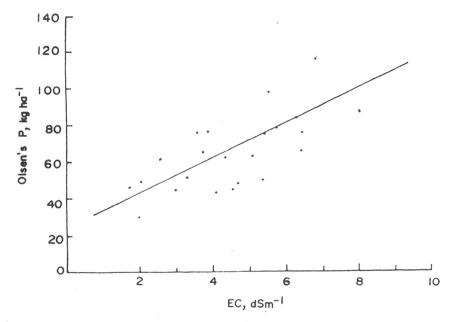

Fig. 5.7.   Relation between EC of 1:2 soil-water suspension and Olsen's P in alkali soils (Chhabra *et al.*, 1980b).

during early stages of reclamation and under conditions where farmers add less amendments resulting in high ESP, soil will keep releasing P in the solution for utilisation by the growing plants (Chhabra, 1988b). For these reasons, there is a positive correlation between soil ESP and Olsen's extractable P (Fig. 5.9). Pratt and Thorne (1948) observed increased P solubility with increase in pH in a Na-dominated clay system, while in Ca-dominated system it decreased. Kaziev (1956) reported that phosphates are more soluble in solonetz Sierozems than in non-saline Sierozems. Kumarswamy and Mosi (1969) reported a negative relationship between soil pH and P fixation capacity leading to higher solubility of P in alkali soils.

Phosphorus extracted by all the three commonly used methods—Olsen's, Bray's, and Kanwar's—gives high correlation with pH and ESP of the soil (Table 5.7). In alkali soils, Ca-P is the dominant inorganic P fraction followed by Al-P, Fe-P, and saloid bound P. The regression analysis revealed that increasing soil pH and ESP significantly increased Al-P and saloid bound P (Chhabra, 1976b). On the basis of the pH and H_2PO_4 relationship in the solubility diagram, Singh *et al.* (1975) reported that octa calcium phosphate is the dominant mineral in alkali soils. The high regression coefficient ($r = 0.75$) between phosphate potential ($pH_2PO_4 + 1/2pCa$) and the lime potential ($pH - 1/2pCa$) further supported the dominance of this mineral in alkali soils.

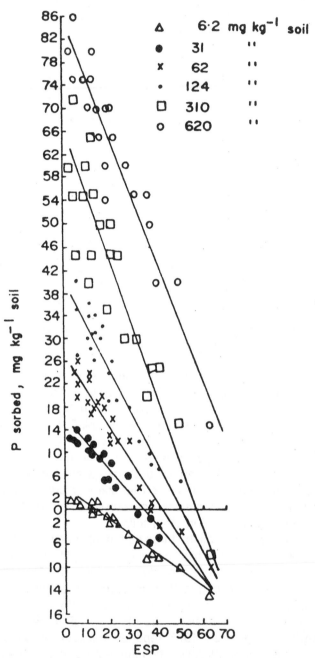

Fig. 5.8.   Sorption of P as affected by soil ESP and concentration of applied phosphorus (Chhabra, 1988b).

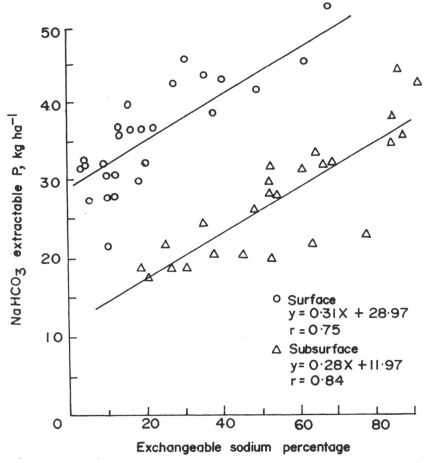

Fig. 5.9. Relation between soil ESP and NaHCO₃ extractable phosphorus of cultivated fields (Chhabra, 1988b).

Table 5.7. Soil phosphorus as extracted by different methods and its relationship with ESP and pH of alkali soils

Method	Range of extractable P, kg/ha, in		r values for surface 15 cm soil	
	0–15 cm soil	15–30 cm soil	ESP	pH
Olsen's	22–46 (34)*	18–44 (28)	0.75**	0.72
Bray's	24–48 (36)	14–46 (24)	0.76	0.71
EDTA-I (pH 4.7)	28–134 (68)	8–44 (24)	0.58	0.58
EDTA-II (pH 8.5)	42–128 (82)	18–68 (36)	0.81	0.76

* Figures in parenthesis are mean values.
** Significant at $P = 0.01$.
Source: Chhabra (1976b).

However, during reclamation of alkali soils, the extractable P of the surface layers decreases. Chhabra *et al.* (1981) attributed this to:

a) immobilisation of soluble sodium phosphate through its conversion into less soluble calcium phosphate by added gypsum,

b) decrease in ESP/pH of the soil leading to greater sorption of soluble P,

c) movement of P from the surface to the lower layers, and

d) removal of soil P by the growing plants.

In a column study (Fig. 5.10), P losses up to 90 mg P per column amounting to 60 ppm P or 134 kg P/ha were observed when gypsum was surface applied

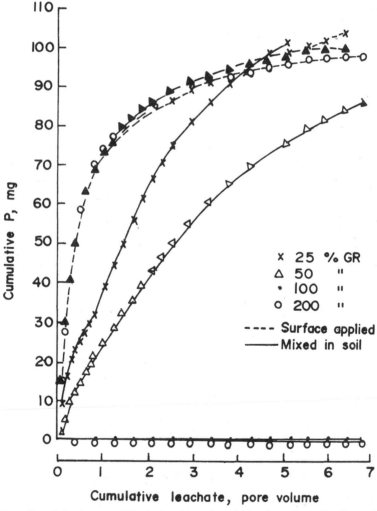

Fig. 5.10. Cumulative losses of P in the leachate of an alkali soil as affected by the method and rate of gypsum application (Chhabra *et al.*, 1981).

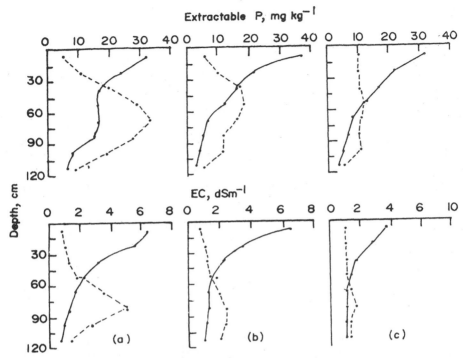

Fig. 5.11. Extractable P and conductivity of an alkali soil profile as affected by gypsum application and leaching (a = 0, b = 5, and c = 10 t/ha; solid lines = before leaching, broken lines = after leaching) (Chhabra *et al.*, 1981).

and the soil leached with water. The losses could be checked when gypsum was mixed in the soil at 100 and 200 per cent of the laboratory-determined gypsum requirement. The above results of controlled laboratory investigations were confirmed by field studies (Fig. 5.11), which revealed that when alkali soils were leached after surface application of gypsum, P along with other soluble salts moved down from the surface to the lower layers. The P distribution pattern and EC profile before and after the leaching showed that P moved along with other salts though the peak P concentration somewhat lagged behind that of other salts. Higher doses of gypsum resulted in higher infiltration rates and deeper penetration of soluble salts including phosphorus. From a long-term field experiment, Chhabra (1985) reported substantial leaching losses of soluble P from the surface 15 cm layer, losses being higher in the initial years. In nature, the leaching losses of soluble P from the rootzone are partly prevented by the poor permeability of the soil and immobilisation by added gypsum.

For the above reasons, rice and wheat (Table 5.8) grown in freshly reclaimed alkali soils did not respond to application of phosphatic fertilisers in the initial 3 to 5 years (Chhabra *et al.* 1980 a, Chhabra and Abrol, 1981; Chhabra, 1985).

After 5 years of continuous cropping, the available P in the surface 15 cm soil of P control plots, R_0W_0 (Fig. 5.12), was depleted but the lower layers continued to have medium to high amounts of extractable P (Chhabra, 1987a). It was observed that if the Olsen's P of the surface 15 cm soil falls below 7.5 kg P/ha, the rice yields may suffer (Chhabra, 1987a). Rice, being a shallow-rooted crop, depends upon the fertility status of the surface 30 cm layer (Fig. 5.13), which becomes depleted for reasons explained earlier and thus starts responding to P fertilisation while wheat plants, which have a relatively

Table 5.8.   Effect of N and P application on the yield, t/ha, of rice and wheat in a gypsum-amended alkali soil

Treatment	Years after reclamation										
	1	2	3	4	5	6	7	8	9	10	11
RICE											
$R_0 W_0$	6.58	7.24	6.65	7.31	5.90	5.89	5.27	3.68	4.27	3.88	5.03
$R_{11}W_0$	7.29	6.89	6.81	7.45	6.31	6.29	6.57	5.13	5.82	5.47	5.52
$R_{22}W_0$	6.78	7.08	7.18	8.24	7.30	7.01	7.48	5.70	5.83	6.05	6.92
R_0W_{11}	6.87	7.32	6.87	8.02	6.92	6.73	6.89	4.12	4.84	4.76	5.57
R_0W_{22}	6.11	7.23	6.92	8.04	6.74	6.66	7.04	4.64	5.48	5.34	6.41
$R_{11}W_{11}$	6.91	7.09	6.95	8.32	6.93	7.30	7.86	5.76	5.48	5.86	6.80
$R_{22}W_{22}$	6.69	7.02	6.83	7.97	6.75	6.82	7.74	5.98	6.17	5.90	7.59
$N_0P_0K_0$	3.94	3.39	2.43	2.67	2.69	2.63	2.11	1.77	2.30	2.59	2.14
LSD at P = 0.05 for											
P levels	NS	NS	NS	0.60	0.55	0.74	1.00	0.76	0.85	0.94	0.76
WHEAT											
$R_0 W_0$	4.35	3.71	3.63	5.04	4.74	4.83	5.55	4.88	5.66	4.32	4.35
$R_{11}W_0$	4.62	3.99	3.46	4.99	4.70	5.04	5.48	5.03	5.62	4.44	4.53
$R_{22}W_0$	4.38	4.02	3.72	5.25	4.86	5.04	5.85	5.37	5.98	4.57	4.55
R_0W_{11}	4.25	3.80	3.62	5.10	4.75	4.88	5.56	5.33	5.55	4.44	4.29
R_0W_{22}	4.34	3.82	3.53	5.17	4.79	4.93	5.49	5.19	5.85	4.74	4.39
$R_{11}W_{11}$	4.49	3.86	3.63	5.27	4.87	4.97	5.74	5.33	5.90	4.69	4.51
$R_{22}W_{22}$	4.45	3.87	3.45	5.16	4.59	4.80	5.51	5.17	5.84	4.83	4.41
$N_0P_0K_0$	0.83	0.85	0.50	0.89	0.92	1.13	1.01	0.91	1.07	0.95	0.82
LSD at P = 0.05 for											
P levels	NS	NS	NS	NS	NS	NS	NS	NS	NS	NS	NS

Source: Chhabra (1987a).

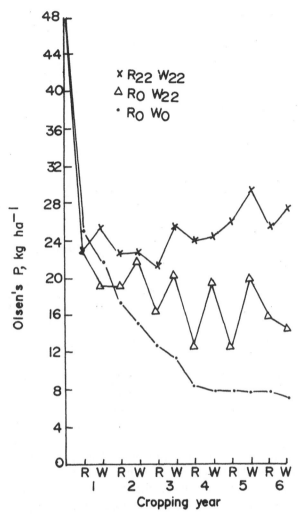

Fig. 5.12.   Changes in Olsen's extractable P of surface 15 cm of a gypsum-amended alkali soil as affected by rice-wheat cropping and P fertilisation (R = rice, W = wheat, 0 and 22 refer to kg P applied per ha) (Chhabra, 1985).

deeper rooting system and can forage P from the lower layers to meet their needs, do not respond to applied P for another 3 to 5 years. In addition to that, considering the growth period, wheat as compared to rice crop gets about 30 days more to exploit soil P and thus to meet its P requirements. For these reasons, not only the yield but also the P content of the plant is more affected in rice than in wheat.

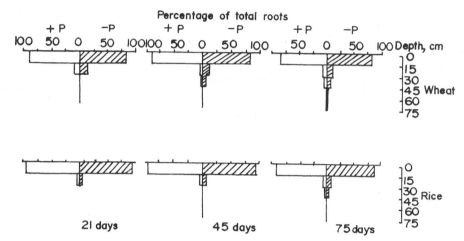

Fig. 5.13.   Effect of P application to rice and wheat on per cent root distribution at different stages of growth in an alkali soil (Chhabra, 1987a)

These results thus point out that, unlike in non-alkali waterlogged soils, P fertilisation should be applied to rice rather than to wheat in a rice-wheat sequence followed in alkali soils. In non-alkali soils, depletion of extractable P by the growing plant from the surface layer is compensated by mobilisation of native soil P, mainly from Fe-P, as a result of waterlogging maintained during rice cultivation. But in alkali soils, which have high pH, low organic matter, and calcium phosphate as being the dominant inorganic P fraction, waterlogging does not play a significant role in mobilising the native soil phosphorus. Furthermore, in view of the contribution from the sub-surface layers, lower doses of P, i.e., 11 kg P/ha, are enough to get optimum yields of rice and wheat crops in these soils. Thus, in alkali soils, proper evaluation of fertility status and accounting for the role of subsoil fertility can help in judicious use of P fertilisers and reduce the cost of crop production in the initial years of reclamation.

Though rice is an efficient P absorber, its uptake is severely inhibited at pH above 7. Using ^{32}P, Chhabra *et al.* (1986b) reported that depletion of P by rice from a nutrient solution was maximum at pH 5 and almost zero at pH 8 and above (Fig. 5.14). This is attributed to the fact that at high pH, as in the alkali soils, P is present in PO_4^{3-} ions, which are physiologically unavailable to the plants. But in soils, pH, of the rhizosphere is low, between 6.2 and 7.2, favouring conversion of trivalent into di- and monovalent forms (HPO_4^{2-} and $H_2PO_4^-$) and thus facilitating their uptake. Thus, physiological availability of phosphates in alkali soils may seldom be the limiting factor in plant growth, provided phosphorus solubility is held at a high level (Chhabra and Abrol, 1983).

When the crops grown in reclaimed alkali soils start responding to P fertilisation, then single superphosphate (SSP) and diammonium phosphate

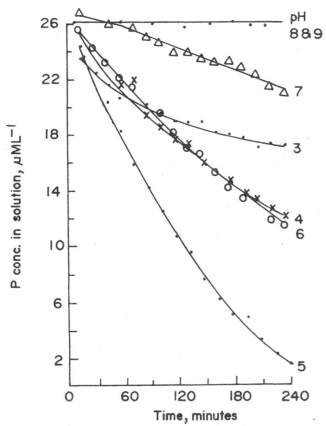

Fig. 5.14.   Effect of pH on the P depletion by rice from nutrient solution (Chhabra *et al.*, 1986b).

(DAP) are better sources than nitrophosphate (NP) to meet the P needs of the crops (Fig. 5.15). Considering the calcareous nature of these soils, fertilizers containing water-soluble P are more effective than those containing water-insoluble phosphorus. Though alkali soils contain high amounts of calcium phosphate (apatite), yet efforts to solubilise and thus make it available through inoculation with P-solubilising bacteria such as *Thiobacillus* have proved unsuccessful (Chhabra, 1986). This is mainly due to the non-functioning of these bacteria at alkaline pH. Hence, to get optimum crop production on a sustainable basis, proper fertilisation must be done in reclaimed alkali soils.

5.1.4. Potassium

In general, increasing soil ESP decreases the K and increases the Na content of the plants (Fig. 5.16). Due to high Na and deficiency of Ca, many studies

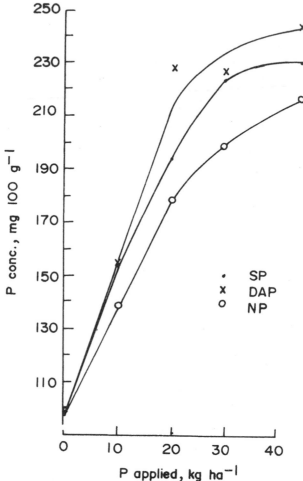

Fig. 5.15.   P content of 21-day-old rice plants as affected by the source and level of phosphorus (Chhabra, 1986).

have shown reduced uptake of K by plants raised in alkali soils. (Martin and Bingham, 1954; Chhabra and Abrol, 1975; Chhabra et al., 1979; Singh *et al.*, 1979, 1980, 1981). However, its absolute concentration in the plant tissue is nearly always above the lower critical limit. In the field, crop responses to applied K fertilisers have not been observed, possibly because of the presence of micaceous minerals and illite (Kanwar, 1961; Sidhu and Gilkes, 1977), which are capable of releasing sufficient K to meet the crop needs. Pal and Mondal (1980) observed an enormous release of K from an alluvial alkali soil with both water and ammonium acetate. Chhabra (1985) reported that barren alluvial alkali soils contain high amounts of exchangeable K, up to 400 kg K/ha. On

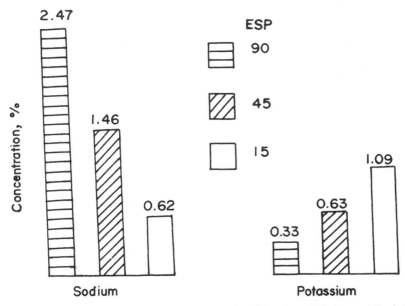

Fig. 5.16.   Effect of ESP levels on chemical composition of rice straw (Chhabra and Abrol, 1975).

cropping such lands, application of K fertilisers increased neither the K content of plants nor yield of rice and wheat. However, the soil exchangeable K in the treatments receiving no K but P and N showed a greater depletion than those receiving N alone (Fig. 5.17). In spite of that, not only the surface but also the sub-surface layers kept having a high amount of exchangeable K to meet the plant requirement (Fig. 5.18).

This was mainly due to replenishment of absorbed K (by the plant) from the exchange sites by the K from non-exchangeable sources. The response to K nutrition also depends upon the nature of the cultivar and more precisely on its capacity to selectively absorb K under strong antagonistic effect of sodium. Joshi *et al.* (1980) reported that tolerant wheat cultivars accumulated more K and less Na than the Na-sensitive varieties and suggested that Na:K ratio may be used as index for prediction of varietal sensitivity to high Na concentration. For correcting Na-induced K deficiency in plants grown in alkali soils with medium to high amounts of exchangeable K, instead of applying K fertilisers, it is suggested that a proper amount of amendment should be added so as to correct Ca:Na balance, which in turn improves the K status of the plants (Fig. 5.19).

5.1.5. Zinc

Alkali soils contain medium to high amounts of total zinc, 40 to 100 mg Zn/kg soil, which is comparable with the amount found in non-alkali soils of the area.

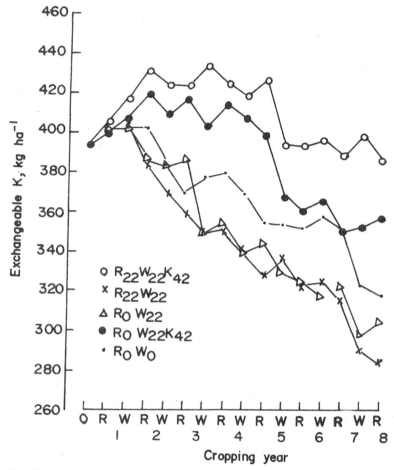

Fig. 5.17. Exchangeable K status of a gypsum-amended alkali soil as affected by differential P and K application over a period of 8 years following rice-wheat cropping sequence (R = rice, W = wheat, K_{42} = potassium applied to wheat only; 42 kg K/ha, 0 and 22 refer to kg P applied per ha) (Chhabra, 1985).

But the availability of Zn to the plants grown in alkali soils is adversely affected by high pH, presence of $CaCO_3$, high soluble P, and low organic matter. Most often, these soils contain less than 0.6 mg DTPA-extractable Zn/kg soil (Katyal *et al.*, 1980; Singh *et al.*, 1984) and show acute deficiency of Zn. A highly significant negative correlation was observed between extractable Zn and pH and $CaCO_3$ content of the soil (Bhumbla and Dhingra, 1964; Kanwar and Randhawa, 1974; Mishra and Pandey, 1976).

The solubility of Zn in alkali soils is governed by the solubility of $Zn(OH)_2$ and $ZnCO_3$, which are the immediate reaction products (Dhillon *et al.*, 1975).

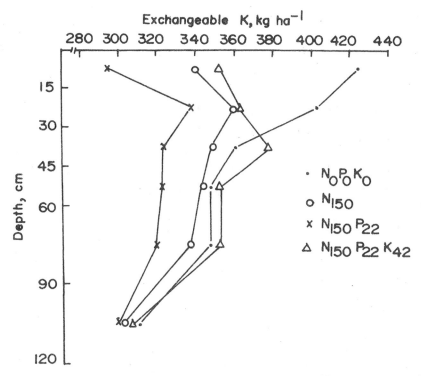

Fig. 5.18. Exchangeable K of an alkali soil as affected by application of fertilisers over a period of 8 years following rice-wheat cropping sequence (Chhabra, unpublished).

Takkar and Sidhu (1979) reported that Zn concentration in the soil solution was regulated by both $Zn(OH)_2$-Zn^{2+} (aq) and $ZnCO_3$-Zn^{2+} (aq) systems during the initial periods (up to 21 days) and thereafter by $ZnCO_3$-Zn^{2+} (aq) system alone because of the buffering effect of soil carbonate equilibria. They observed the value of pZn + 2 pOH to vary between 16 and 17.5 and that of pZn + pCO_3 between 17.6 to 18.8.

The solubility and extractability of added Zn decreases with time but increases with increase in ESP of the soil (Singh *et al.*, 1984). The higher Zn extractability at high ESP is attributed to the formation of sodium zincate, which is soluble. On addition of amendments the extractability of added Zn decreases (Milap Chand *et al.*, 1980; Singh *et al.*, 1984). This is due to:

a) greater adsorption of Zn by Ca- than by Na-saturated soil,

b) retention of added Zn on the surface of freshly precipitated $CaCO_3$, formed as a result of reaction between soluble carbonates and added gypsum, and

c) enhanced competition of added Ca with Zn for the DTPA ligands during extraction.

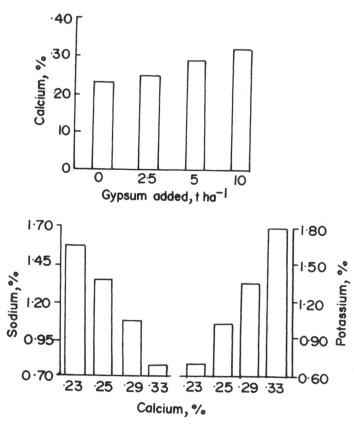

Fig. 5.19.   Effect of gypsum application on Ca concentration and its influence on Na and K concentration of rice in an alkali soil (Chhabra and Abrol, 1987).

Formation of calcium zincate complex during reclamation of alkali soils is ruled out as the concentration of Zn ions and the pH required for its formation, as studied in the pure system, are very high. Shukla *et al.* (1980) observed that Zn adsorption in soils saturated with various cations was in the order of H < Ca < Mg < K < Na. The exchange of Zn with different ions was non-spontaneous and significantly influenced by the pH of the suspension.

Rice crop, though tolerant to soil sodicity, is sensitive to Zn deficiency, which may appear 15 to 21 days after transplanting in the form of rusty-brown spots (Fig. 5.20) on fully matured leaves causing stunted growth, poor tillering, and low yields. In extreme cases the whole leaf becomes dark brown and the field gives a burned appearance. The commonly known "Khaira" disease in most rice-growing areas is related to Zn deficiency. In marginal cases, the plants recover after 30 to 45 days and put up new leaves and tillers, but maturity is

Fig. 5.20. Zinc deficiency symptoms shown by paddy grown in an alkali soil (Chhabra, un-published).

delayed by 7 to 10 days. Moderate to severe yield reduction due to Zn deficiency have been reported by several workers (Kanwar and Randhawa, 1974; Chhabra, 1976a; Takkar and Randhawa, 1978; Singh et al., 1979). Besides the addition of amendment and nitrogen, application of Zn is important for optimum crop yields in alkali soils (Fig. 5.21). From field studies, Singh et al. (1987) reported that in soils amended with 10 to 15 t/ha of gypsum, 10 to 20 kg $ZnSO_4$/ha was enough to meet Zn requirement of rice and wheat (Table 5.9). At low level of gypsum application and moderate dose of applied Zn, the plants suffer more from Na toxicity and Ca deficiency than from Zn availability. It was also observed that higher application of $ZnSO_4$ increased the DTPA-Zn of the soil (Fig. 5.22) and Zn content of plants, but did not increase the grain yield further. Chhabra et al. (1982) observed that after 4 to 5 years of continuous zinc application in a reclaimed alkali soil, the available Zn status of the soil increased to 1.6 mg DTPA extractable Zn/kg soil. Under such conditions, further application of $ZnSO_4$ can be deferred for 3 to 4 years without any loss in grain yield.

Application of $ZnSO_4$ at the time of sowing or transplanting is the best way to meet Zn needs of crops grown in alkali soils. From extensive field studies at the Central Soil Salinity Research Institute, Karnal, India, for 8 years, conduted by the author, revealed that zincated urea is a good source of Zn for alkali soils. Since both N and Zn are deficient in alkali soils, such a complex fertiliser is highly beneficial for these soils. Application of Zn-enriched FYM, heavy application of Zn in nurseries, and two to three sprays of 0.5 per cent $ZnSO_4$ solution have been found to correct Zn deficiency. Dipping of seedling roots in 3 per cent suspension of ZnO, before transplanting is equally beneficial (Chhabra et al. 1988).

Table 5.9.   Effect of gypsum and zinc sulphate on the yield, t/ha, of rice in an alkali soil

Zn SO_4, kg/ha	Gypsum level, t/ha				Mean
	0	2.5	5	10	
0	0.33	0.98	2.14	2.65	1.48
10	0.49	1.87	3.06	3.77	2.30
20	0.58	2.00	3.14	3.85	2.39
30	0.68	1.75	2.99	3.92	2.34
40	1.05	2.02	3.29	3.89	2.56
Mean	0.59	1.72	·2.93	3.62	

LSD at P = 0.05, gypsum, = 3.6,   interaction = NS, $ZnSO_4$ = 2.8.
Source: Singh et al. (1987).

Fig. 5.21.   Effect of zinc application on the growth of rice in a gypsum-amended alkali soil (Chhabra, unpublished).

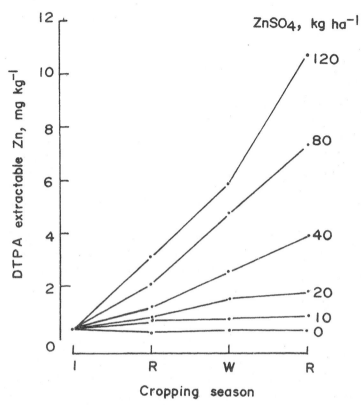

Fig. 5.22. Effect of differential zinc sulphate, kg/ha, application on the DTPA-extractable Zn in a rice-wheat cropping sequence in an alkali soil (I = initial, R & W = after rice and wheat) (Singh *et al.*, 1987).

5.1.6. Iron

Next to zinc, iron is the most common deficient micronutrient in alkali soils. Hale and Wallace (1960) reported that high amounts of CO_3^{2-} and HCO_3^-, as is the case in alkali soils, lead to decrease in Fe accumulation in soybean plants and ultimately to HCO_3^--induced Fe chlorosis. Severe damage has been observed due to induced chlorosis in rice nurseries grown in alkali soils and on raised beds in calcareous soils (Singh and Singh, 1975; Takkar and Randhawa, 1978; Katyal and Sharma, 1980). Iron chlorosis appearing as white strips on the leaf along its length is often observed in sugarcane grown on calcareous and alkali soils.

Since iron solubility is conditioned by pH, $CaCO_3$, oxidation status of the soil, and amount of organic matter, it is not the total Fe but the available iron that is a limiting factor in alkali soils. Correction of iron deficiency using soluble salts such as $FeSO_4$ is generally not found useful unless it is accompanied by changes in the oxidation status of the soil brought about by prolonged submergence and addition of easily decomposable organic matter (Shahi *et al.*, 1976; Katyal and Sharma, 1980). Swarup (1980) showed a marked increase in the extractable Fe and Mn status of an alkali soil upon submergence up to 60 days. The increase was greater when organic materials such as rice husk and FYM were incorporated in the soil. Pretreatment of nursery beds after incorporation of green manure, FYM, or compost helps in creating reduced conditions and mobilising soil iron to meet the plant needs. Hence, it is always advisable to apply excess water to rice nurseries when they show iron deficiency symptoms. Pyrites (FeS_2) can be applied in non-alkali calcareous soils to augment Fe supply to the plants.

Since Fe is an immobile nutrient in the plant, foliar application of Fe (3 per cent solution of $FeSO_4$) gives a limited relief to the suffering crop and should be used to supplement the improvement in reduction status of the soil.

5.1.7. Manganese

The solubility and availability of Mn in soil is also governed by pH and oxidation-reduction status of the soil. Bhatnagar *et al.* (1966) reported that though total Mn was greater in alkali soils of Chambal command area of India, the active and ratio of active to total Mn was less than that of normal soil. Singh (1970) observed that in most of the soils, exchangeable and active Mn was negatively correlated with pH and $CaCO_3$. Pasricha and Ponnamperuma (1976) found that the effect of soluble NaCl and $NaHCO_3$ on the kinetics of water-soluble Mn were similar to those of Fe^{2+}. Salts like $NaHCO_3$ depress its availability. Deficiency of Mn is seldom a problem for wetland rice, while it can become a serious limiting factor for crops like wheat following it, because, on submergence, Mn gets reduced and solubilises earlier than iron and is leached to the lower layers. As a result, Mn deficiency is increasingly being observed in wheat grown in rice-wheat cropping sequence on coarse-textured alkali soils. Mn deficiency symptoms in wheat come after first irrigation, i.e., at the stage of maximum tillering, in the form of white spots in the centre of the leaf blade. As a consequence, the leaf breaks from that point and dies. At maturity, Mn-deficient plants experience difficulty in ear emergence and proper grain formation.

Due to auto-oxidation of Mn, it is very difficult to correct Mn deficiency by soil application of $MnSO_4$. Repeated sprays of $MnSO_4$ are needed to make up the deficiency of this element in upland crops. Deep ploughing so as to mix subsoil with the surface layers generally helps in redistribution of soil Mn and increases its availability to the plants.

5.1.8. Boron

Boron is not likely to be deficient in alkali soils (Bhumbla and Chhabra, 1982). In fact, more often it is likely to be present in toxic amounts as its availability increases with increase in soil pH/ESP. Kanwar and Singh (1961) observed a positive correlation between water-soluble B and the pH and EC of soils. However, Bhumbla *et al.* (1980) in a survey of uncultivated alkali soils of Punjab and Haryana (India) observed that though the surface 15 cm soil contained high amounts of B soluble in hot water, up to 25 mg/kg soil, there was no significant correlation between B, pH, and EC of the soil. Singh and Randhawa (1977) reported that water-soluble B in saline and alkali soils of Punjab constituted 3.8 to 21 per cent of total B. Tourmaline is the major source of B in these soils, though illite also contributes a little. They further observed that 25 per cent of total B was present as leachable boron.

Bandyopadhya (1974) observed a seasonal variation in the B content of alkali soils (pH and ESP varying between 9.6 to 10.7 and 85 to 100, respectively), due to highly soluble nature. During the dry season, B accumulated in the surface 15 cm soil, while during the rainy season it got leached down along with other salts and accumulated in the lower depths.

In addition to leaching, B hazards in alkali soils can be minimised by addition of gypsum. Gupta and Chandra (1972) from a laboratory study observed a marked reduction in water-soluble B together with pH and SAR, on addition of gypsum to a highly alkali soil. At high pH/ESP, boron is present as highly soluble sodium metaborate, which upon addition of gypsum is converted into relatively insoluble calcium metaborate. The solubility of calcium metaborate is very low, 0.4 per cent as compared with 26 to 30 per cent of sodium metaborate. For these reasons, in spite of their initial high B content, crops grown in gypsum-amended alkali soils do not suffer from B toxicity. Ashok Kumar and Abrol (1982) reported reduced uptake of B even by grasses grown in gypsum-amended alkali soils.

5.1.9. Molybdenum

Like that of B, the solubility of Mo increases with increase in pH. Pasricha and Randhawa (1971) reported that recently reclaimed alkali soils of Sangrur district, India, contained oxalic acid–extractable Mo from 0.012 to 0.440 mg/kg soil in the surface 15 cm and from 0.063 to 0.720 mg/kg soil in sub-surface (15–30 cm) soil. Higher levels of extractable Mo in the sub-surface layers in these soils is due to its higher solubility under impeded drainage conditions. Since the extractable Mo in these soils is greater than the critical limit of 0.15 mg/kg soil, fodder grown on these soils is likely to accumulate Mo in excess quantities that may prove toxic to the animals feeding on it. Application of adequate amounts of gypsum decreases the Mo, content of the plants through the antagonistic effect of SO_4^{2-} on MoO_4^{2-} absorption by the roots.

5.1.10. Fluorine

The solubility of fluorine, like that of B and Mo, is affected by pH and sodicity of the soil. Chhabra *et al.* (1980c) studied the effect of varying levels of ESP/pH on the solubility and adsorption of fluorine, an element more important to animal health. Water-extractable F (Chhabra *et al.*, 1980b) increases with increase in soil ESP and pH (Fig 5.23), the latter having a more important role in determining the behaviour of F in soils (Gupta *et al.*, 1982). Adsorption of F in alkali soils, as described by the Langmuir isotherm (Chhabra *et al.*, 1980c), showed that at any equilibrium F concentration in the soil, there was a decrease in F adsorption with increase in soil ESP/pH (Fig. 5.24). Addition of F-containing fertilisers such as superphosphate, rockphosphate, and nitrophosphate and amendments such as phosphogypsum, which may contain F from traces to 12 per cent, may increase the water-soluble F if the pH/ESP of the soil remains high. A high amount of F in alkali soils decreases the rice yields (Singh *et al.* 1979a), disturbs Na:Ca ratio by decreasing Ca but increasing Na content of the plants, and increases the F content of rice (Fig. 5.25) and wheat straw (Fig. 5.26), thus endangering the health of the animals that may feed on them. Toxic effect of F can be minimised by lowering the ESP/pH by addition of amendments and by antagonistic effect of P on the absorption of F by plant roots (Singh *et al.*, 1979a, 1979b).

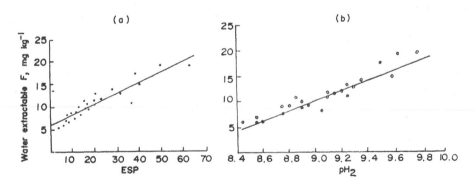

Fig. 5.23.   Water-extractable F as affected by (a) ESP and (b) pH_2 in alkali soils (Chhabra *et al.*, 1980c).

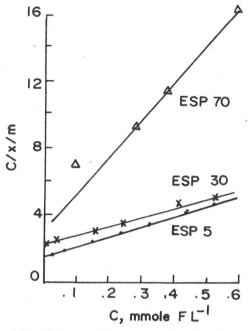

Fig. 5.24. Adsorption of F as affected by ESP of the soil (Chhabra *et al.*, 1980c).

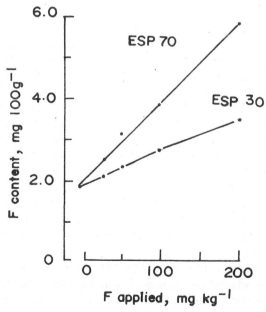

Fig. 5.25. F content of rice straw as affected by soil ESP and applied fluorine (Singh *et al.*, 1979a).

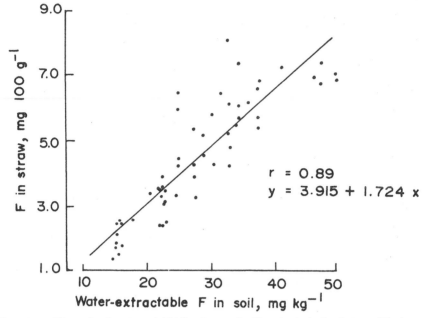

Fig. 5.26.   Effect of soil water-soluble fluorine on the F content of wheat straw (Singh *et al.*, 1979b).

5.2. Saline Soils

In contrast to alkali soils, saline soils contain an excess of neutral soluble salts, mostly Cl^- and SO_4^{2-} of Na^+, Ca^{2+}, and Mg^{2+}, and pH of the saturated soil paste is less than 8.2. Poor plant growth in saline soils is due to the high osmotic pressure of soil solution causing low physiological availability of water, direct toxic effect of individual ions, and a complex interaction between Na, Ca, and K resulting in the disturbed equilibrium of these elements in the plant. Poor aeration due to high water table and poor quality irrigation water may further restrict the plant's ability to absorb nutrients in required amounts.

In addition to other agronomic practices, as discussed earlier, harmful effects of low to moderate salinity can be alleviated by judicious use of fertilisers or plant nutrients. It has been observed that moderate soil salinity may even interact positively with the plant nutrients and enhance the metabolism resulting in high dry matter production (Dregne and Mojallali, 1969; Hassan *et al.*, 1970).

For most crop plants, yield decreases with increase in salinity but for a given salinity level the yields can be increased by application of fertilisers. Bernstein *et al.* (1974) have given three possible types of interaction between salinity and fertility level of the soil (Fig. 5.27). In Fig. 5.27a, response to salinity is proportionally (but not arithmetically) the same at both levels of fertility,

Yield

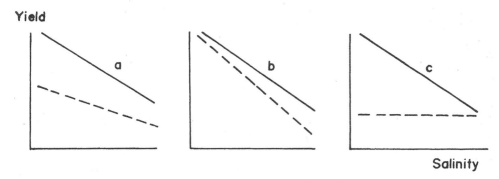

Salinity

Fig. 5.27. Salinity fertility interaction: salinity effect (a) independent of fertility level, (b) greater at low fertility, and (c) greater at high fertility level (—— high fertility, ---- low fertility) (Bernstein *et al.*, 1974).

meaning that, in general, there will be increase in yield due to increased fertility at all the salinity levels. This is the situation when there is absolute deficiency of the nutrient. Under such a situation, there is no significant interaction between salinity and fertility of the soil. In Fig. 5.27b, the effect of salinity is low at high fertility level, indicating that at high salinity harmful effects can be minimised by increasing the fertility status of the soil. In Fig 5.27c, the salinity effect is greater at the higher fertility level, indicating that salts used to increase fertility may themselves have an additive effect on the adverse response of salinity.

Responses to added fertilisers (Table 5.10) are generally observed at low to medium levels of salinity, normally below 8 to 10 dS/m (Luken, 1962; Dregne and Mojallali, 1969; Hassan *et al.*, 1970); at higher salinity levels, the yields are poor because of the adverse effect of salinity itself.

Table 5.10. Effect of salinity and nitrogen levels on the yield , g/pot, of wheat

Salinity levels ECe, dS/m	Nitrogen levels, mg N/kg soil				Mean
	0	80	120	160	
1.3	5.8	16.8	22.2	26.7	17.9
8.2	14.4	27.2	28.2	30.5	25.3
15.9	3.5	6.3	8.2	10.1	7.3
17.9	3.1	1.0	1.8	0.7	0.7
Mean	7.4	12.8	15.1	17.0	

Source: Al-Rawi and Sadallah (1980).

Excess soluble salts in the root environment have a confounding influence on the mineral nutrition of plants. Hassan *et al.* (1970) observed that in saline soils dominated by Cl^- and SO_4^{2-} of Na^+, Ca^{2+}, and Mg^{2+}, barley plants showed increased uptake of Na, Mg, Mn, and Zn but decreased uptake of P, K, Ca, and Fe. These nutritional disorders could be expected to accentuate the yield limitation imposed by osmotic effects of salinity.

Quite often it is believed that addition of inorganic fertilisers may aggravate the problem caused by excess salts. Lunin and Gallatin (1965) even observed an increase in salt status of the soil with increased application of fertilisers. In practice, however, inherent lack of essential nutrients is responsible for reduced productivity of many saline soils. Also, when saline soils are reclaimed by leaching with large quantities of water, some of the essential plant nutrients are lost in the process. This further necessitates the application of fertilisers in adequate amounts to get satisfactory yields.

The salinity affects the nutrient availability by:
a)   modifying the retention, fixation, and transformation of the nutrients in soils,
b)   interfering with the uptake and/or absorption of nutrients by roots due to ionic competition and reduced root growth,
c)   disturbing the metabolism of nutrients within the plants, mainly through water stress, and thus reducing their effectiveness.

The behaviour of different nutrient elements in saline soils is discussed below.

5.2.1. Nitrogen

Some of the important factors that govern N fertilisation of crops grown in saline soils are:
a)   Inherent deficiency of available nitrogen in saline soils (Sharma *et al.,* 1968; Singh *et al.,* 1969; 1983).
b)   High leaching losses of nitrogen as NO_3^- (Torres and Bingham, 1973).
c)   Decreased rate of nitrification of ammonia due to high salinity and direct toxic effects of Cl^- on the bacterial activity. Agarwal *et al.* (1971) reported that chloride salts depressed nitrification, whereas low concentrations of SO_4^{2-} promoted it. Westerman and Tucker (1974) observed that high concentration of salts (KCl and K_2SO_4) inhibited nitrification and caused NH_4^+-N to accumulate. McClang and Frankenberger (1985) reported that in saline soils, though the hydrolysis of urea and thus the production of ammonia was not affected, its nitrification was severely inhibited. Thus, higher concentration of salts inhibit nitrification, resulting in reduced production of NO_3^--N. Hence, plants that can absorb N only as NO_3^- will show a deficiency of N even though it may be present in the soil. Therefore, it is better to use NO_3^--N than NH_4^+-N fertilisers in saline soils.
d)   Reduced uptake of NO_3^- due to antagonistic effect of Cl^- and SO_4^{2-} (Torres and Bingham, 1973). However, total N uptake by plants is

generally not affected by the level of salinity. Khalil *et al.* (1967) observed that N percentage in corn plants increased with N application as well as with salinity level. Thus, the plant continued to accumulate N under saline conditions in spite of reduction in the dry matter yield. Yield of N (dry matter × per cent N) showed a slight but statistically insignificant decrease with increase in salinity. Reduction in N fertiliser efficiency, therefore, was primarily due to retardation of plant growth under saline conditions rather than due to reduction in N uptake. Langdale and Thomas (1971) observed that N content of barley plants increased with increase in the rate of nitrogen application at all levels of salinity, indicating that N absorption by plant was not upset by salinity.

However, in waterlogged saline soils, poor aeration and anaerobic conditions may restrict the availability to and absorption of N by plants. It has been observed that for a given level of applied N, the NO_3^- content in soil was lower under high water table than under low water table. This was attributed to greater reduction of NO_3^- to NO_2^- under anaerobic conditions. Woodruff *et al.* (1984) further concluded that poor drainage associated with a high water table led to N deficiency and that maize crop raised on such soils required more fertiliser N for producing maximum yield (Fig. 5.28). Similarly, after heavy rainfall leading to water stagnation and anaerobic conditions, plants show yellowing, a characteristic deficiency symptom of N, and need an extra dose of fertiliser to offset the adverse effect.

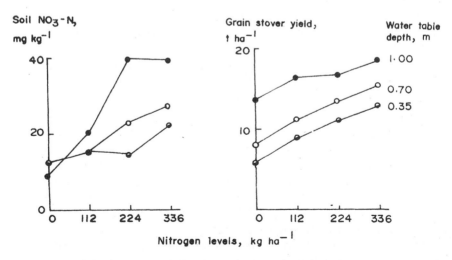

Fig. 5.28.   Soil NO_3^- -N and grain stover yield of maize as affected by rate of N fertilisation and water table depth (Woodruff *et al.*, 1984).

e) Poor symbiotic N fixation due to toxic effects of salts on *Rhizobia* leading to drastic reduction in nodulation (Grahm and Parker, 1964; Bernstein and Ogata, 1966; Yadav and Vyas, 1971; Subha Rao *et al.,* 1972; Lakshmi Kumari *et al.,* 1974; Kumar and Garg, 1980). Grahm and Parker (1964) observed that normal *Rhizobia* associated with pea can tolerate a maximum salinity of 4.5 dS/m.

f) Reduced N metabolism within the plant due to water stress brought about by high osmotic pressure of the soil solution. Nightingale and Farnham (1936) reported that with increase in the osmotic pressure, the amount of soluble organic N and protein content of sweet pea roots decreased, while the NO_3^- -N accumulated, showing that protein synthesis was inhibited under high salinity conditions. Wadleigh and Ayers (1945) observed that water and salt stress reduce the capacity of red bean plants to incorporate and metabolise L-leucine-^{14}C into protein. They also observed accumulation of NO_3^- -N in plants under increased salinity. Similar results were reported by Ben-Zioni *et al.* (1967) and Kahane and Mayber (1968), who concluded that water deficiency caused by salt stress was responsible for reduced metabolism in the plants grown in saline media. Thus, it would appear that the total N status of plant is a poor index of salinity hazards. Indeed, the protein-N should be the true index of N status of plants grown under saline environment. In salinity-tolerant plants, protein synthesis is relatively unaffected. Bhola *et al.* (1980) observed that in *Kharchia* and *Kalyan Sona* wheat cultivars, salinity up to 16 dS/m did not affect the protein content.

Responses to applied N in saline soils are, therefore, due to their poor N status caused by the above factors. Luken (1962) reported that wheat yields increased by application of N and NP under low to moderate levels of salinity. Amer *et al.* (1964) reported that application of N increased cotton yield in a saline soil (ECe 1.7 dS/m) in Egypt. Khalil *et al.* (1967) observed that yield of corn decreased with salinity (ECe ranging from 1.6 to 9.0 dS/m) but increased with N application.

Generally, in saline soils, responses to applied N have been observed at lower levels of application than in alkali and normal soils. Bhattacharya (1972) observed that highest yield of rice in coastal saline soils of Sunderban (India) was obtained by an addition of 40 kg N and 9 kg P/ha. Sharma (1980) observed that fruit yield of eggplants grown in saline soils of the Yemen Arab Republic increased on application of 40 kg N/ha and higher doses up to 120 kg N/ha did not further improve the yield significantly.

Not only the amount but also the method, time of application, and source of N affects its efficiency in saline soils. Split application of N in saline soils is as effective as in normal and alkali soils. In coastal saline soils (Biswas, 1975),

Table 5.11. Effect of splitting and time of N application on the yield of rice in a saline soil with high water table

Nitrogen proportion			Grain yield[*], t/ha	
Basal	Tillering	Flowering	Summer rice (*Kharif*)	Winter rice (*Rabi*)
0	0	0	2.75	2.75
75	25	0	4.25	3.40
0	25	75	4.60	4.00
0	100	0	4.10	3.80
50	50	0	4.20	3.10
0	50	50	4.85	4.15
25	75	0	4.15	3.15
0	75	25	5.40	5 45
33	34	33	5.60	4.90

[*]Average of two years.
Source: Biswas (1975).

omitting basal application of N and applying it in two splits, i.e., at 45 and 60 days after transplanting, gave better results (Table 5.11). On inland saline soils, top dressing of N should be done after and not before irrigation, because salinity in the rootzone increases between the two irrigations. The increased salts, just before the irrigation, prevent its nitrification and availability to the plants.

Foliar application of N (3 per cent solution of urea, 20 kg N/ha) along with soil application is very economical and beneficial in saline soils. This also saves one irrigation (a definite advantage in areas with poor quality water) that otherwise is to be given to dissolve and effectively distribute the top-dressed fertiliser. Sharma (1980) observed that placement of fertiliser (75 per cent) accompanied by foliar spray (25 per cent) gave better results than broadcasting and 100 per cent placement. Among the different N sources, urea gives better results than CAN and $(NH_4)_2SO_4$ under low to medium salinity. But at higher salinity, it is better to use NO_3^- rather than NH_4^+-containing fertilisers.

In certain parts of the world, e.g., Haryana, Rajasthan, Uttar Pradesh, Madhya Pradesh and Andhra Pradesh in India, some parts of California, and some parts of Pakistan and Egypt, the underground saline waters being used for irrigation contain appreciable amounts of NO_3^-. Though these can meet the N requirements of crops, they sometimes become toxic to the plants. Continuous irrigation with such waters pushes up vegetative growth, delays maturity, and affects grain filling adversely. Normally, the grains are shrivelled and are of poor quality. For such areas, the last two irrigations should be given by non-saline canal waters and, instead of grain crops, fodder crops should be grown.

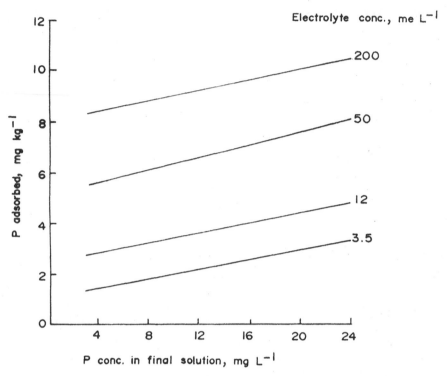

Fig. 5.29.   Effect of electrolyte concentration on the adsorption of P by soil (El-Mahi and Mustafa, 1980).

5.2.2. Phosphorus

Availability of phosphorus to the plants is modified by the soil salinity or saline irrigation water for the following reasons:

a) Precipitation of applied phosphorus
Taylor and Gurney (1965) observed that KCl and NH_4NO_3 precipitated about 15 to 25 per cent of soluble P of the $Ca(H_2PO_4)_2 \cdot 2H_2O$ *in situ* and thus decreased the availability of added phosphorus. Balba (1980a) observed enhanced immobilisation of P in superphosphate applied to an Egyptian soil in the presence of NaCl and ascribed that to the precipitation through increased Ca solubility from native $CaCO_3$ in the presence of NaCl.

b)Higher retention of soluble P by soil
Ryden and Syers (1975) reported that with increase in ionic strength of the soil solution, as is the case in saline soils, retention of P increased. El-Mahi

and Mustafa (1980) reported that saline soils retain more P (Fig. 5.29) than non-saline soils and thus will need more P fertilisation to get good crops. However, Khalil *et al.* (1967) using ^{32}P observed no change in availability of soil phosphorus by increasing soil salinity. They also did not observe any change in specific availability of fertiliser phosphorus by increasing soil salinity. Contrary to that, Janardhan and Rao (1970) observed a decrease in the per cent utilisation of fertiliser P by rice plant with increase in soil salinity.

c) Anion competition
As phosphates and chlorides, both being anions, are absorbed by essentially the same mechanism, excess concentration of Cl⁻ as found in highly saline soils may severely affect the uptake of phosphates because of competitive inhibition (Zhukovskaya, 1973; Chhabra *et al.*, 1976; Manchanda and Sharma, 1983). As a result, per cent P content and total P uptake by the plant are generally affected by soil salinity. Lunin and Gallatin (1965) observed that P content of bean leaves decreased with increasing salinity, especially at highest salinity level of 6 dS/m. Dahiya and Singh (1976) reported that concentration and uptake of P by pea greatly decreased with increase in both EC and ESP, salinity being more harmful than alkalinity. Chhabra *et al.* (1986b) observed a significant decrease in the rate of P depletion by rice as the salinity of the nutrient solution increased from 5 to 250 (Fig. 5.30).

d) Reduced root growth
Phosphorus is a relatively immobile nutrient in saline soils and for its uptake the plant roots must mine the soil. Increasing soil salinity reduces root growth, which in turn decreases the surface area of the roots in contact with P in the soil and thus results in reduced uptake of phosphorus (Khalil *et al.*, 1967; Hassan *et al.*, 1970).

As a consequence, generally P application increases crop yields in saline soils. Fine and Carson (1954) observed that on a moderately saline soil, symptoms regarded as due to salt injury, e.g., leaf-tip necrosis and reddish colour of lower leaves of small grain crops (barley and oats), were also associated with P deficiency. Application of large quantities of phosphatic fertilisers, in both greenhouse and the field, increased yields and alleviated those symptoms. Ferguson and Herlin (1963) reported that much higher responses to applied P occurred on moderately saline soil than on a non-saline soil of comparable available P status (Table 5.12). Patel and Wallace (1975) observed an increase in the yield of tomato fruits, fresh corn cobs, and green foliage of Sudan grass by application of P to a saline soil. Like that of N, response of applied P is limited to low and moderate salinity levels. Luken (1962) and Dregne and Mojallali (1969) observed that beneficial effect of applied P to wheat and barley, respectively, was limited up to an ECe of 9 dS/m.

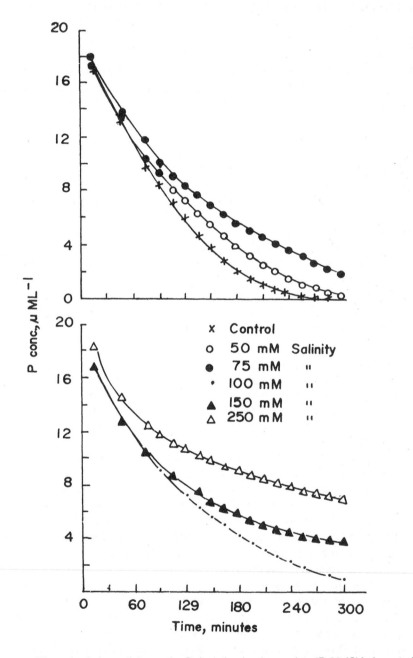

Fig. 5.30. Effect of solution salinity on the P depletion by rice, variety IR-28 (Chhabra *et al.*, 1986b).

Table 5.12. Effect of P application on the relative yield of wheat grown in saline and non-saline soils

P levels, kg P/ha	Per cent yield	
	Non-saline	Saline
0	100	100
4.2	113	132
8.4	117	143
16.8	123	152
LSD at P = 0.5	NS	41.5
NaHCO$_3$-extractable P, kg P/ha	22.2	22.6
ECe, dS/m	0.6	6.2
pHs	7.6	7.7

Source: Ferguson and Herlin (1963).

The above results show that higher plant responses to applied P occur on moderately saline than on non-saline soils. It may be cautioned that responses to applied P fertilisers in saline soils cannot be explained on the basis of soil test values alone, as the saline soils even when containing high amounts of extractable P have shown positive responses to applied phosphorus. This is because, in saline soils, the availability of P is more a function of plant root length and area (which is restricted because of salinity) and the antagonistic effect of excess chloride on P absorption by roots.

To sum up, application of phosphatic fertilisers in saline soils helps in increasing the crop yields by directly providing phosphorus and by decreasing the absorption of toxic elements such as Cl$^-$ (Chhabra *et al.*, 1976) and F$^-$ (Singh *et al.*, 1979b). Thus, while application of P fertilisers may be missed in the initial years of alkali soil reclamation, their addition is a must in saline soils.

5.2.3. Potassium

Saline soils quite often contain medium to high amounts of available potassium (Sharma *et al.*, 1968), but during reclamation through leaching, losses of K may take place, leading to its low inherent availability (Torres and Bingham, 1973). Generally, plants grown under high salinity show K deficiency due to antagonistic effect of Na and Ca on K absorption and/or disturbed Na:K or Ca:K ratio. Under such conditions, application of potassic fertilisers may increase plant yields.

Sodium toxicity may be less common in saline soils than in alkali soils. However, excess sodium may cause a reduction of K within the plant, leading to low yields. El-Elgabaly (1955) working with resin-sand mixtures observed that increasing Na concentration decreased K content of barley plants and induced thick, short, brown, unbranched roots. Balba (1960) showed that K uptake by plants was depressed by irrigation with water containing NaCl. Finck (1977) found that wheat grown on slightly to moderate saline soils in southern Italy suffered from excess of Na and deficiency of Ca and potassium.

Not only the absolute content of K but also its balance with respect to Na, Ca, and Mg may be affected by soil salinity, resulting in reduced crop yields. El-Elgabaly (1955) observed that in plants grown in saline soils, Ca may accumulate in high concentrations, resulting in unfavourable Ca:K ratio. In a greenhouse study, Francois and Bernstein (1964) observed a decrease in K and Mg contents, while there was an increase in Ca, Na, and Cl contents of safflower. Their results further showed that, though the K content decreased with increase in salinity, the relationship between Ca:K ratio and yield was more significant than with K alone, showing thereby that disturbed Ca:K ratio may be a more important factor limiting yields under high salinity conditions. Thomas (1980) reported that with an increase in salinity, Na, Ca, and Mg contents of cotton plants increased but the concentration of K was not affected. However, the disturbed K:(Ca + Mg) ratio became a limiting factor for cotton yield.

On moderately saline soils, application of potassic fertilisers may increase the crop yields (Dregne and Mojallali, 1969; Thomas, 1980) either by directly supplying K or by improving its balance with respect to Na, Ca, and Mg. However, under high salinity conditions it is difficult to effectively exclude Na from the plant by use of K fertilisers.

5.2.4. Micronutrients

Very little is known about the trace element nutrition of crops grown in saline soils. Reports of micronutrient (Fe, Mn, Cu, and Zn) deficiencies in saline soils are rather uncommon.

Often, there is a decrease in the micronutrient content of plants with an increase in salinity. Hassan *et al.* (1970) reported that Fe and Cu contents of corn and barley decreased, while those of Mn and Zn increased with increase in salinity. They observed that, under SO_4^{2-} type of salinity, precipitation of Fe as $Fe_2(SO_4)_3$ may cause low availability to plants. Maas *et al.* (1972) concluded from the solution culture studies that the concentration of Mn and Fe in the vegetative parts of tomato, soybean, and squash decreased with increase in level of salinity but even in the highest salinity treatment, the concentration of these nutrients remained within the physiological limits necessary for plant growth. Similarly, Patel and Wallace (1975) observed a decrease in Cu, Fe, Mn, Si, and Mo and sometimes Zn concentration in plant tissues of tomato, sweet corn, and Sudan grass by an increase in soil salinity and attributed this to high Ca content of the saline soils. However, they also concluded that such a decrease in micronutrient levels would not result in deficiencies.

5.2.5. Fertiliser Application and Cl⁻ Toxicity

In many cases, the adverse effects of high soil salinity can in fact be traced to the toxic effects of high concentration of chlorides in the soil solution. Chloride toxicity is characterised by leaf burn, starting at the tip of the older leaves and progressing along the leaf margin. In many woody species it may also lead to early leaf drop and defoliation (Ayers and Westcot, 1985). High concentration of Cl⁻ disturbs amylolytic activity, resulting in accumulation of starch. It also interferes with the normal photosynthetic activity of protoplasm and, causes reduction in chlorophyll content and carbohydrate synthesis even though there may be apparent accumulation of sugars and starch in the leaves (Wadleigh and Ayers, 1945). Noggle (1966) and Thomas and Langdale (1980) observed that Cl⁻ decreased the organic anions concentration in several plant species and that may be responsible for growth retardation at high salinity.

Proper application of N and P fertilisers can decrease the toxic concentration of soil absorbed Cl⁻ in the plants. Dregne and Mojallali (1969) reported that increasing P application had a moderately depressing effect on chloride uptake. Chhabra *et al.* (1976) using ^{32}P and ^{36}Cl observed that application of

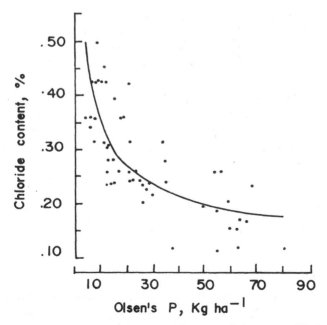

Fig. 5.31.   Effect of soil available phosphorus on Cl⁻ content of wheat straw (Singh *et al.* 1979b).

P drastically reduced the rate of ^{36}Cl uptake by intact tomato plants. From the enzyme kinetics they observed that though both these anions are absorbed by the same mechanism, it has higher affinity for P than for Cl⁻ and thus, under adequate supply of P in the medium, uptake of Cl⁻ is highly reduced. Singh *et al.* (1979b) reported significant decrease in per cent chloride content of wheat straw (Fig. 5.31) with increase in soil available phosphorus. Similarly, application of N decreased Cl⁻ concentration in the plants (Sameni *et al.,* 1980). Thomas and Langdale (1980) observed that unfavourable balance of organic anions, caused by excess Cl⁻, can be corrected by application of nitrogen. Thus, application of a proper amount of fertilisers, especially of N and P, helps alleviate Cl⁻ toxicity in saline soils.

5.2.6. Fertiliser Application and Water Use Efficiency

Plant growth in saline soils is controlled largely by the physiological availability of water and toxic effect of specific ions. Reduced entry of water into roots caused by salinity or water stress inhibits the meristematic activity and elongation of the root, and later on, the development of secondary tissues. Vegetative growth is, therefore, retarded as the salinity builds up or the osmotic pressure increases.

Proper application of fertilisers enables better root growth and sometimes helps in its deeper penetration, which enables the plant to absorb more water (from the lower layers) and thus increases crop production. Kelley (1951) observed that application of fertilisers does not alter evapotranspiration (ET) but increases the yield (Y), thereby increasing the water-use efficiency (ET/Y).

Application of adequate amounts of nutrients may also help shorten the vegetative growth period and hasten maturity, thus minimising the number of irrigations required for completing the life cycle of the plant. The amount and time of application of N fertilisers can be adjusted so as to hasten the maturity or at least avert unnecessary delay in maturity. Application of P and K in deficient soils and/or under high salinity conditions can avert delay in maturity of wheat, oats, soybean, corn, and grapes. Zinc fertilisation in deficient soils hastens the maturity of beans, corn, and rice.

Hence, the balanced use of fertilisers can help in better utilisation of soil moisture, increase the water-use efficiency, and minimise the total irrigation requirement of the crops and thus help save water, availability of which is the most important constraint in saline soils.

5.3. Post-reclamation Management of Fertilisers

Management of fertility in salt-affected soils is important not only in the initial years of reclamation but also in the later stages. It has been commonly observed that after 3 to 4 years of reclamation and continuous harvest of good crops, the farmers complain of decreased yields and conclude that the soil has deteriorated again. Such a situation should be carefully analysed. It may just happen that when the soil starts giving normal yields and/or the extension worker goes away, the farmers become less attentive and do not apply the recommended doses of fertilisers. At the same time, because of increased crop production, the surface soil may get depleted of one or more nutrients. So a soil sample must be tested and recommended amounts of nutrients added on the basis of the test so as to sustain good yields.

In view of the specific needs of salt-affected soils, general application of multinutrient fertilisers such as NPK mixtures, zincated superphosphate, and diammonium phosphate may not always be necessary in the initial years of reclamation. Avoiding their use will save the money and minimise the cost of reclamation. However, after reclamation, basal application of the complex fertilisers should be encouraged for balanced nutrition of the crops.

6

Irrigation Water: Quality Criteria

Water is one of the most important inputs for realising and sustaining high agricultural production. However, its management (transportation to the farm, method and frequency of irrigation) and quality are intimately related to the development of waterlogging and soil salinity.

In most cases, excess soluble salts, leading to soil salinisation, may be attributed to the weathering products accumulated in the soil (primary salinisation). However, under arid and semi-arid conditions, irrigation water may be instrumental in accumulation of salts in the rootzone that were originally equally distributed in the soil profile or localised in deeper layers, thus causing development of soil salinity. This phenomenon, known as secondary salinisations, is the major cause of decreased production on introduction of irrigation in many parts of the world.

Under certain situations, such as near the sea coast or where underground water is used for irrigation, water itself can be a source of salts and lead to development of soil salinity. In general, the assessment of water quality criteria is based on the consideration of two related aspects, i.e., the possible effects on the physico-chemical properties of the soil and the impact on the crop yield. The main criteria for assessing the quality of water for irrigation are salinity, sodicity hazards and specific ion effects.

6.1. Salinity hazards

The most important criterion regarding salinity and therefore of water availability to the plant is the total salt concentration. Total salt concentration in water can be measured in terms of milligram per litre (mg/l), parts per million (ppm), per cent (%), and milliequivalent per litre (me/l). Since there exists a straight line correlation between electrical conductivity (EC) and total salt concentration of waters (Fig. 6.1), just like in dilute solutions, the most expedient procedure to evaluate salinity hazards is to measure its electrical conductivity (as in soil extracts). The common unit for expressing EC values

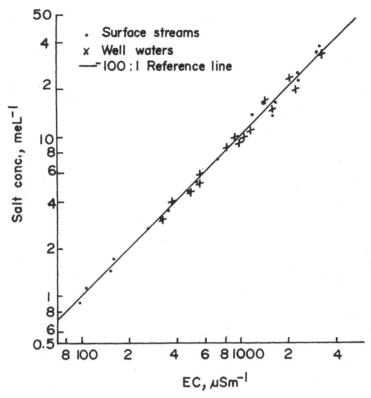

Fig. 6.1. Concentration-electrical conductivity relationship for irrigation waters (Richards, 1954).

of irrigation waters is micromhos per centimetre (μ mhos/cm) or microSiemens per metre (μ S/m) in contrast to dS/m, used for soil extracts. From the EC values, the total salt concentration can be calculated using the following formula, equation (1):

$$\text{Salt concentration, me/l} = \frac{\text{EC}(\mu S/m)}{100} \tag{1}$$

Many times, salt concentration is also expressed as total dissolved solids (TDS), which are determined gravimetrically. TDS can also be calculated from the EC of the water using the following equation (2):

$$\text{TDS, mg/l} = \text{EC (dS/m)} \times 640 \tag{2}$$

On the basis of salt concentration, US Salinity Laboratory Staff (Richards, 1954) divided the irrigation waters into four classes. Later on, another class was added to it (Table 6.1). Most of the irrigation waters fall into the range of

Table 6.1.   Salinity hazards of irrigation waters

EC of irrigation waters, μ S/m	Salinity class	Salinity hazards
100–250	C_1	very low
250–750	C_2	low
750–2250	C_3	medium
2250–5000	C_4	high
>5000	C_5	very high

Source: Richards (1954).

150 to 1500 μ S/m. Though waters having EC values above 1500 μ S/m can cause serious damage, waters having EC values as high as 8000 μ S/m are being used on light-textured soils having deep water table without any harmful effect on soil or plant.

In general the EC values of the SP extracts of soils are significantly higher than those of the applied water, usually by a factor of 2 to 10, depending on soil depth, frequency of irrigation and amount of water applied. When water is applied to the soil, much of it is removed through evapotranspiration, leaving the bulk of soluble salts behind. As a result, salt concentration increases proportionally (in the absence of precipitation) to the decrease in volume of water left in the soil. When a new increment of water is applied to a (partly) water-depleted soil, it displaces most of the pre-existing soil solution downward, taking the place of the previous solution, a phenomenon known as miscible displacement.

Therefore, the corresponding volume of displaced solution contains higher concentrations of dissolved salts than present in the applied water. Consequently, the salt concentration in the rootzone increases as the leaching fraction (i.e., the fraction of the applied water passing through the rootzone) decreases. Although these aspects will be dealt with quantitatively at a later stage, it is important to stress at this point that, whatever may be the salt concentration of the irrigation water applied, some excess water (above the evapotranspiration needs of the crop) is mandatory to leach out the salts. Evidently, the higher the salt concentration, the higher is the leaching requirement. It is this process just described that is responsible for the variation of EC in the soil solution with time and depth in the profile. Usually (i.e., under good management conditions), the concentration of salts in the topsoil, say down to a depth of 20 to 30 cm, is about 2 to 3 times that of the irrigation water. It may, however, increase by a factor of 5 to 10 at greater depths.

6.2. Sodicity Hazards

In addition to total salinity, the tendency of irrigation water to generate excessive levels of exchangeable sodium, which adversely affect the soil physicochemical properties, needs to be considered. The useful parameters for expressing the sodium hazards of irrigation waters are (a) sodium adsorption ratio and (b) residual sodium carbonate.

6.2.1. Sodium Adsorption Ratio

Sodium adsorption ratio (SAR) of the water is calculated from equation (3) as described by Richards (1954).

$$\text{SAR} = \frac{Na^+}{\sqrt{\dfrac{Ca^{2+} + Mg^{2+}}{2}}} \tag{3}$$

where Na^+, Ca^{2+}, and Mg^{2+} are in me/l.

From the SAR of the irrigation water, ESR and ESP of the soil can be calculated using the following equation:

$$\frac{\text{ESP}}{100} = \frac{\text{ESR}}{1 + \text{ESR}} = \frac{(-\,0.0126 + 0.01475\,\text{SAR})}{1 + (-\,0.0126 + 0.01475\,\text{SAR})} \tag{4}$$

On the basis of SAR, the irrigation waters have been divided into four categories (Table 6.2).

Table 6.2. Sodicity hazards of irrigation waters

SAR of irrigation water, me/l	Sodicity class	Sodicity hazards
< 10	S_1	Low
10–18	S_2	Medium
18–26	S_3	High
> 26	S_4	Very high

Source: Richards (1954).

From the field studies, it has been observed that the values of ESP actually obtained are generally higher than those expected on the basis of SAR of the irrigation waters. This is due to the fact that in the soil, concentration of salts in the irrigation water increases (as a result of evapotranspiration). Therefore, the SAR of the water also increases but not with the same magnitude. For example, if both Na^+ and $Ca^{2+} + Mg^{2+}$ concentrations increase by the factor of 2, the SAR increases by a factor of $\sqrt{2}$ (Table 6.3).

Table 6.3. Effect of salt concentration on the SAR of irrigation water

Conc. of cations, me/l			Total cations, me/l	SAR, me/l
Na^+	Ca^{2+}	Mg^{2+}		
4	2	2	8	2.83
8	4	4	16	4.00
12	6	6	24	4.90

However, it should be stressed that the above reasoning applies at the steady state, i.e., when the soil has reached equilibrium with the irrigation water (sometimes referred to as "yield value"). Depending on the CEC of the soil, the amount of water applied, and the salt concentration in the irrigation water, the time required to reach this equilibrium may vary from a few to 10 years or more. The relation between the (steady-state) ESP and the SAR value of the applied irrigation water is shown in Fig. 6.2. The solid line in the figure represents the relation based on equation (3) and is also indicated by the scales C and D in the nomogram shown in Fig. 3.11, Chapter 3. In the same figure, a dotted curve shows the ESP values that would be attained if the total salt concentration of the irrigation water were to increase by a factor of 3 (or

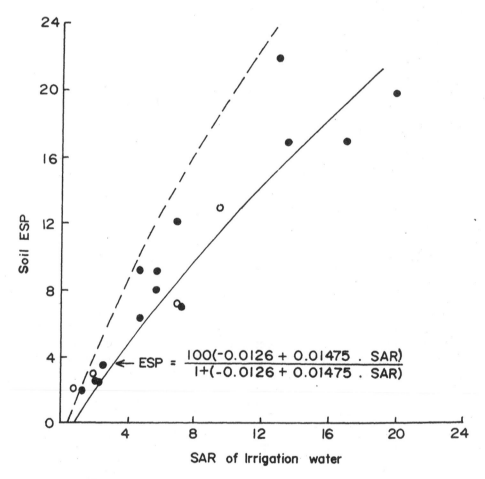

$$ESP = \frac{100(-0.0126 + 0.01475 \cdot SAR)}{1 + (-0.0126 + 0.01475 \cdot SAR)}$$

Fig. 6.2.   ESP dependence on SAR of irrigation waters from lysimeter (o) and field observations (●) (Richards, 1954).

SAR by a factor of 1.7). The distribution of the experimental points tends to indicate that the behaviour of most of these soils is consistent with a "concentration effect" by a factor of 2 to 3. In spite of these difficulties, it would nevertheless appear that the SAR value is a very useful index for assessing the sodium hazard of the irrigation water.

As emphasised before, we are here concerned with the discussion of the steady-state values. However, in the initial stages of irrigation, it is very unlikely that, in regard to balance between Na^+ and $Ca^{2+} + Mg^{2+}$, soil and irrigation water are in equilibrium. If, in the initial stages, ESR values of the soil are quite low and SAR values of the irrigation water are high, the time required to achieve equilibrium may be quite long, particularly if the CEC is high and the salt concentration of the irrigation water is low. This stems from the fact that the overall soil composition is well buffered against changes in ESR since the total ionic pool in the exchange complex is considerably larger than the amount of cations introduced in any one irrigation treatment.

It is obvious that the buffering power of the soil is dependent on both the CEC of the soil and the salt concentration of the irrigation water. These principles can be illustrated by a simple batch test (involving no change in total salt concentration as a result of water loss). Let us take the following example: 1 kg of a soil (dry, salt-free) is "dispersed" in 1 litre of irrigation water of a total salt concentration of 10 me/l (9 me Na^+, 1 me $Ca^{2+} + Mg^{2+}$) and a SAR value of 12.7. The ESR value of the soil is 0.02. Evidently, these two phases are far from being in equilibrium and a readjustment of ion composition is bound to occur. This will correspond to a shift of Na^+ ions from the irrigation water to the soil, displacing an equivalent amount of $Ca^{2+} + Mg^{2+}$ ions (from the soil) in the process, until an equilibrium is achieved. We take it that the equilibrium condition corresponds to the requirement ESR = 0.015 SAR. We may discuss two cases corresponding to CEC values of (a) 20 and (b) 5 me/100 g soil, respectively.

a)   CEC = 20 me/100 g; or Na^+ = 4 me/kg soil

$$Ca^{2+} + Mg^{2+} = 196 \text{ me/kg soil}$$

This equilibrium condition is described by the equation (5):

$$\frac{4 + x}{196 - x} = 0.015 \frac{9 - x}{\sqrt{\dfrac{1 + x}{2}}} \tag{5}$$

in which x represents the amount of Na^+ replacing an equivalent amount of $Ca^{2+} + Mg^{2+}$ from the exchange complex. The solution of this equation looks rather cumbersome and is best achieved by calculating the left- and right-hand side for a range of x values. The results are then plotted on a graph (Fig. 6.3) and the answer is provided by the intersection point of two curves.

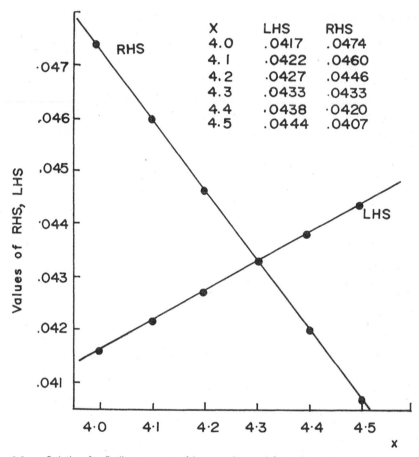

X	LHS	RHS
4.0	.0417	.0474
4.1	.0422	.0460
4.2	.0427	.0446
4.3	.0433	.0433
4.4	.0438	.0420
4.5	.0444	.0407

Fig. 6.3.   Solution for finding amount of ions exchanged from the soil and irrigation water (Richards, 1954).

Alternatively, x may be found (by trial and error) until the answer is consistent with equation (5). In this particular case, the value of x comes to 4.30. In other words, 4.30 me of Na^+ are shifted into the exchange complex, replacing an equivalent amount of $Ca^{2+} + Mg^{2+}$. The overall result is an increase in soil ESR from 0.02 to 0.044, whereas the SAR of the water has decreased from 12.7 to 2.9.

b)   Now let us take the case of a second soil where:

$CEC = 5$ me/100 g; or Na^+ = 1 me/kg soil

$Ca^{2+} + Mg^{2+}$ = 49 me/kg soil for ESR of
0.02 me/100 g soil

The equilibrium condition in this case is described by the following equation:

$$\frac{4 + x}{49 - x} = 0.015 \, \frac{\dfrac{9 - x}{\sqrt{1 + x}}}{2} \tag{6}$$

In this soil the value of x at equilibrium comes to 2.45. Putting this value in equation (4) and recalculating the concentrations of Na^+ and $Ca^{2+} + Mg^{2+}$, it is observed that ESR of the soil has increased from 0.02 to 0.074, while the SAR of the water has decreased from 12.7 to 5.0. From this example, it appears that the buffering power of the soil (against ESR variations) is directly related to CEC values. Similarly, the relative effect on soil ESR from an irrigation water of identical SAR value but lower salt concentration would be less pronounced. Moreover, it appears that the relative effect of subsequent irrigation treatments on soil ESR would decrease, i.e., the ESR of the soil would increase more rapidly in the initial stages and approach asymptotically towards the yield value demanded by the SAR value of the irrigation water.

6.2.2. Residual Sodium Carbonate

One of the characteristics of many ground and river waters is the presence of bicarbonates. In irrigation waters containing high concentrations of bicarbonate, there is a tendency to precipitate Ca^{2+} and Mg^{2+} (to a lesser extent) as carbonates when the soil solution becomes more concentrated. Evidently, the depletion of bivalent ions in the soil solution leads to an increase *in situ* of SAR and consequently to ESP values larger than those that could be anticipated on the basis of SAR values of the irrigation waters (Eaton, 1950).

The effect of the presence of bicarbonate on ESP is illustrated in Table 6.4, which shows the results of a "low leaching" irrigation treatment of a pot-grown Rhodes grass in a Hanford loam soil. The treatment involved irrigation with the waters shown, allowing the soil to dry to a moisture tension of 700–800 cm between intermittent irrigations. At every fourth irrigation, excess irrigation water was applied so that 25 per cent of the applied irrigation water could be collected as percolate. Soil samples were collected and analysed after the 42nd and 86th irrigations. A series of comments can be made with regard to these data:

— In all cases, the ECe values are significantly higher than EC of the irrigation water, the factor amounting to a range of 2–8, and the effect increases with time.

— In the absence of HCO_3^-, ESP values generally exceed the predictions based on the SAR values of the irrigation water (parenthesis, first column). However, a "square root correction" on SAR (i.e., SAR multiplied by the square root of the ratio of ECe/EC$_{IW}$) predicts ESP values reasonably well (parenthesis, ESP data).

Table 6.4.   Bicarbonate effects on soil ESP

Treatment	SAR	Composition of irrigation water, me/l					Properties of the soil					
							After 42nd irrigation			After 86th irrigation		
		Ca^{2+}	Na^+	HCO_3^-	Cl^-	RSC	EC_e	pH	ESP	EC_e	pH	ESP
1	9.5	5.00	15.00		20.0	0	3.7	6.8	12.0 (13.0)	5.30	6.8	16.0 (16.5)
1'	(ESP = 11)	5.00	15.00	10.0	10.0	5.0	5.9	8.6	52.0	16.00	9.4	72.0
2	6.7	2.50	7.50		10.0	0	2.2	7.2	8.4 (9.5)	3.70	7.3	15.0 (12.9)
2'	(ESP = 8)	2.50	7.50	5.0	5.0	2.5	2.0	8.6	20.0	7.30	9.0	42.0
3	4.7	1.25	3.75		5.0	0	1.3	6.8	9.0 (7.6)	1.80	6.7	11.0 (8.9)
3'	(ESP = 5.4)	1.25	3.75	2.5	2.5	1.25	1.2	8.4	10.0	2.40	7.7	20.0
4	2.1	0.25	0.75		1.0	0	0.4	7.1	2.4	0.32	6.4	2.4
4'	(ESP = 2)	0.25	0.75	0.50	0.5	0.25	0.36	7.0	2.6	0.34	6.4	2.1
5	0.9	3.75	1.25		5.0	0	1.0	6.9	2.2	1.50	6.1	2.2
5'	(ESP = 1)	3.75	1.25	4.25	0.75	0.5	0.70	8.1	3.1	1.10	7.4	3.5
6	0.4	0.75	0.25		1.0	0	0.34	7.0	1.5	0.36	6.4	1.1
6'	(ESP = 0.5)	0.75	0.25	0.85	0.15	0.10	0.33	7.2	1.4	0.34	6.4	1.3

— In the "high" salt concentration irrigation waters, treatments 1, 2, 3, and 5, the presence of HCO_3^- leads to pH values higher than 8; at low salt concentrations, buffering power of the soil ensures a neutral pH.

— In the high salt concentration irrigation waters, the presence of HCO_3^- leads to a significant increase in ESP (at otherwise identical SAR values) that cannot be explained by a salt concentration effect. Moreover, after 86 treatments, ESP values are increased by a factor of about 2.

— At low salt concentrations and low SAR values (treatments 4, 5, 6), the presence of HCO_3^- has practically no effect on ESP.

Various formulations have been proposed to account for the effect of HCO_3^- on sodicity hazards of irrigation water. One of these empirical approaches is based on the assumption that all Ca^{2+} and Mg^{2+} precipitate as carbonates. Considering this hypothesis, Eaton (1950) proposed the concept of residual sodium carbonate (RSC) for the assessment of high carbonate waters.

$$RSC = \left(CO_3^{2-} + HCO_3^-\right) - \left(Ca^{2+} + Mg^{2+}\right) \tag{7}$$

where all ionic concentrations are expressed in me/l.

From practical experience, it appears that RSC > 2.50 is definitely hazardous, 1.25–2.50 is marginal, and < 1.25 me/l is safe. This approach, however, has limited predictive value for estimating ESP and is, in any case, not very realistic because of the assumption of quantitative precipitation.

6.2.3. Adjusted SAR

A definite improvement results from the concept of adjusted SAR (Bower and Maasland, 1963). This is based on the use of the Langelier "saturation index" (SI), originally developed for predicting carbonate deposition in boilers. It relies on a calculation of the pH a given water would arrive at when equilibrated with solid phase $CaCO_3$ at average soil CO_2 levels. This solution-composition–dependent pH, as compared to the initial pH of the water, can be used to predict whether $CaCO_3$ will precipitate or be redissolved by the water, when passing through a calcareous soil (the pH of which is 8.4, when in equilibrium with atmospheric CO_2). The SI is defined as:

$$SI = 8.4 - pHc$$

$$pHc = \{(pK_2 - pK_{sp}) + pCa^{2+} + p(Alk)\} \tag{8}$$

In this equation, pK_2 and pK_{sp} refer to the negative logarithm of the dissociation constant of $H_2CO_3^-$ and the $CaCO_3$ solubility product, pCa is the negative logarithm of Ca^{2+} concentration, and p(Alk) is the negative logarithm of alkalinity concentration ($HCO_3^- + CO_3^{2-}$). Values of pHc below 8.4 indicate the tendency to precipitate lime from the applied water; the opposite effect, i.e., redissolution of lime, occurs at pHc > 8.4. Since the common practice is to include both Ca^{2+} and Mg^{2+}, this expression takes the form of:

$$pHc = \Delta pK + p\left(Ca^{2+} + Mg^{2+}\right) + p\left(HCO_3^- + CO_3^{2-}\right) \qquad (9)$$

ΔpK refers to the difference $pK_2 - pK_{sp}$.

The adjusted SAR is now (Ayers and Westcot, 1976) defined as:

$$SAR_{adj} = SAR_{IW}\left[1 + (8.4 - pHc)\right]. \qquad (10)$$

As will be discussed later, it turns out that in many field situations and in "normal" conditions of irrigation management, the ESP (yield value) in the topsoil is very nearly equal to the adjusted SAR.

$$ESP = SAR_{adj} = SAR_{IW}\left[1 + (8.4 - pHc)\right] \qquad (11)$$

The various parameters used in the calculation of pHc and SAR_{adj} are given in Table 6.5. A series of examples for different compositions of irrigation waters are given in Table 6.6.

There is, however, one additional and important remark to be made. It can be expected that the tendency for $CaCO_3$ precipitation (which is also related to concentration effects in the soil) will also be dependent on the leaching

Table 6.5. Parameters for calculating pHc for various combinations of ionic concentrations

Concentration, me/l	$pK_2 - pK_{sp}$	$p(Ca^{2+} + Mg^{2+})$	$p(Alk)$
0.50	2.10	3.60	3.30
0.75	2.10	3.43	3.12
1.00	2.15	3.30	3.00
1.50	2.15	3.12	2.82
2.00	2.15	3.00	2.70
2.50	2.15	2.90	2.60
3.00	2.20	2.82	2.52
4.00	2.20	2.70	2.40
5.00	2.20	2.60	2.30
6.00	2.20	2.52	2.22
8.00	2.25	2.40	2.10
10.00	2.30	2.30	2.00
12.50	2.30	2.20	1.90
15.00	2.30	2.12	1.82
20.00	2.40	2.00	1.70
25.00	2.40	1.90	1.60
30.00	2.40	1.82	1.52
35.00	2.40	1.76	1.46
40.00	2.45	1.70	1.40
50.00	2.45	1.60	1.30

Source: Ayers and Westcot (1976).

Table 6.6. Examples of compositional combinations in irrigation waters and their characteristic quality parameters

Origin	Ion concentration, me/l							EC, dS/m	SAR	RSC, me/l	pHc	SAR adj
	Na^+	K^+	Ca^{2+}	Mg^{2+}	Cl^-	SO_4^{2-}	HCO_3^-					
Grand River, USA	7.08	0.19	2.0	0.79	0.19	3.43	6.29	0.94	6.00	3.50	7.35	12.30
Colorado River, USA	3.35	0.22	6.95	3.63	1.03	9.31	3.73	1.27	1.45	− 6.80	7.00	3.50
Pecos River, USA	11.38	0.08	16.98	9.07	12.13	22.39	3.11	3.26	3.15	−22.90	6.82	8.30
Nahal-Oz Well, Israel	41.30		1.30	3.70	35.20	1.50	9.60	4.60	26.10	4.60	7.07	61.00

Example: Grand River: $Na^+ + K^+ + Mg^{2+} + Ca^{2+}$ = 10 me/l
$Ca^{2+} + Mg^{2+}$ = 2.79 me/l
HCO_3^- = 6.29 me/l

pK = 2.3; p($Ca^{2+} + Mg^{2+}$) = 2.85; p(HCO_3^-) = 2.20; pHc = 7.35
SAR_{adj} = (1 + (8.4−7.35)) × SAR = 12.3

fraction (or the excess water applied). At very low leaching fraction, the fraction of calcium precipitated can therefore be expected to increase. This is demonstrated in Fig. 6.4 by some experimental data on the effect of pHc on the fraction of carbonate precipitation at varying LF values.

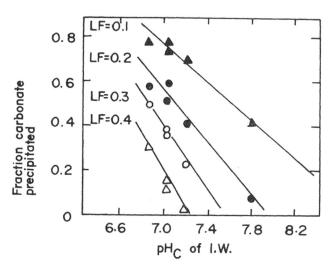

Fig. 6.4.   CaCO₃ precipitation as related to pHc and leaching fraction (Ayers and Westcot, 1976).

It is seen that, as expected, when pHc increases, the $CaCO_3$ precipitation decreases and does not occur at pHc = 8.3 to 8.4, whatever be the LF value. At pHc < 8.4, there is a linear decrease in $CaCO_3$ precipitation with LF values. Moreover, it is apparent that, at any given pHc value, the extent of $CaCO_3$ precipitation decreases at increasing LF values, i.e., when irrigation water is added more copiously. It is of course this effect that is responsible for the excessively high ESP values found in treatments 1 and 2, as reported in Table 6.4, which corresponds to rather low leaching fractions (LF < 0.1).

The effect of LF has tentatively been incorporated in empirical modifications of equation (10) such as

$$ESP = \frac{1}{\sqrt{LF}} \left[SAR_{IW} \left(1 + (8.4 - pH_c) \right) \right] \tag{12}$$

Another version, which is more applicable to the topsoil section, is:

$$ESP = SAR_{IW} \left[1 + (8.4 - pH_c) \right]^2 \tag{13}$$

This equation has some particular relevance because, by expressing a more pronounced sensitivity to bicarbonate effects, it predicts the surface soil

ESP, which is the critical characteristic affecting soil permeability. One final comment is perhaps in order: unless the alkaline pH is caused by excessive bicarbonate levels, the pH of the natural irrigation water is generally not considered as a critical quality parameter on account of the strong buffering properties of soils. However, for the industrial effluents used for irrigation, pH should be one of the critical parameters affecting their suitability.

6.3. Specific Ion Effects

In addition to salinity and sodicity hazards, crops may get affected by low to moderate and high concentration of certain ions that may cause specific toxic symptoms and/or nutritional disorders.

a) Calcium

If the calcium concentration is greater than 35 per cent of the total cations, then the water is fit for irrigation.

b) Sodium

If the sodium concentration is greater than 60 per cent of the total cations, then the water is unfit for irrigation. Excess of Na causes deficiency of K and scorching of leaf-tips in addition to deterioration in soil physical conditions.

c) Potassium

Water containing a high concentration of K is considered good as K alleviates to some extent the harmful effect of sodium.

d) Chlorides and sulphates

Among the anions present in the macro levels, the main concern goes to Cl^- and SO_4^{2-}, the concentration of which is strongly correlated with the EC of the irrigation water. Chloride-sensitive plants may suffer when Cl^- in the SP extract exceeds 5 to 10 me/l. There is, however, a relatively wide range of sensitivity to Na^+ and Cl^- effects among different plant species and even among varieties within a given species. Woody species like deciduous trees, citrus, and avocado are very sensitive to Cl^- toxicity. Annual crops can, however, absorb a relatively high concentration of chloride before showing the specific toxicity symptoms. The sensitivity to Na^+ and Cl^- concentration is significantly enhanced under sprinkler irrigation, where foliar uptake may take place. Fruit plants are particulary sensitive in this respect; concentrations of Cl^- as low as 2 to 3 me/l may cause leaf damage. These effects are particularly pronounced in day-time irrigation and may be alleviated by night time application of irrigation water.

Waters high in SO_4^{2-}, commonly having $Cl^-:SO_4^{2-}$ ratio of 1:3 or higher, and known as SO_4^{2-}-waters, are more deleterious than Cl^- waters because high concentration of SO_4^{2-} is more injurious to roots and disturbs the internal metabolism of the plants. At the same time, continuous use of such waters leads to precipitation of Ca^{2+} as $CaSO_4$, causing rise in pH and ESP of the soil.

e) Boron

Boron has been observed to occur in varying amounts in groundwater of arid and semi-arid zones. The usual range of boron is from 0.1 to 5.0 mg/l. It is of course essential for plant growth but is very toxic at concentration only slightly above optimal levels (i.e., the range between deficiency and toxicity is rather narrow for many crops). Therefore, boron analysis should routinely be included in any appraisal of salinity and quality of irrigation water. When a water of specific boron concentration is used for irrigation, toxicity is relatively less in heavy textured than in light-textured soils because of differential adsorption charac-teristics of boron in the soil (Table 6.7).

Table 6.7.   Boron adsorption, mg/kg, in soils as affected by texture

Dune sand		Sandy loam		Black clay soil	
Equilibrium conc. of B	B adsorbed	Equilibrium conc. of B	B adsorbed	Equilibrium conc. of B	B adsorbed
3.8	1.2	3.0	2.0	1.2	3.8
7.8	2.2	6.4	3.6	3.8	6.2
16.5	3.5	14.8	5.2	8.5	11.5
25.0	5.0	24.0	6.0	14.0	16.0
34.0	6.0	33.0	7.0	14.0	21.0
43.0	7.0	42.2	8.0	25.0	25.0

Source: Gupta (1979).

The currently accepted classification of crops according to boron sensitivity is given in Table 6.8.

f) Lithium

Highly saline groundwaters have been noted to contain appreciable concentra-tions of lithium. The examination of 35 saline groundwater samples from Mathura district of Uttar Pradesh, India, revealed that waters having EC less than 4 dS/m contained lithium less than 0.05 mg/l. Higher salinity waters (EC, 6 to 24 dS/m) contained an average of 0.35 mg/l lithium. The range of lithium varied from 0.02 to 1.9 mg/l, respectively (Table 6.9).

Germination studies with wheat, barley, rice, bengal gram, and maize indicated that increasing concentrations of lithium cause toxic effects. The

Table 6.8. Liimits of born in irrigation water for sensitive, semi-tolerant, and tolerant crops (in descending order of tolerance within the range shown), as based on toxicity symptoms observed in sand cultures

Sensitive 0.3–1 mg/l boron	Semi-tolerant 1–2 mg/l boron	Tolerant, 2–4 mg/l boron
Citrus	Lima bean	Carrot
Avocado	Sweet potato	Lettuce
Apricot	Bell pepper	Cabbage
Peach	Oat	Turnip
Cherry	Milo	Onion
Persimmon	Corn	Broad bean
Fig	Wheat	Alfalfa
Grapes	Barley	Garden beet
Apple	Olive	Mangel
Pear	Field pea	Sugarbeet
Plum	Radish	Palm
Navy bean	Tomato	Asparagus
Jerusalem artichoke	Cotton	
Walnut	Potato	
	Sunflower	

Source: Eaton (1935).

Table 6.9. Lithium in groundwaters of Mathura district, India

ECIW, dS/m	No. of samples	Lithium conc., mg/l	
		Range	Average
1.7 –2.0	6	0.02–0.05	0.04
1.9 –3.6	3	0.04–0.06	0.05
4.0 –5.6	9	0.02–0.14	0.06
6.4–24.8	17	0.05–1.90	0.35

Source: Gupta (1982).

tolerance of different crops varies in the order of wheat > barley > rice > maize > bengal gram. The adverse effect of Li is greater on the roots than on the shoots. The lethal concentration of lithium is much less than 5 mg/l, for sensitive crops such as bengal gram. However, in wheat, the grain yield is not significantly affected up to 20 mg/l of lithium.

g) Magnesium

The saline groundwaters occurring in arid and semi-arid zones commonly contain magnesium higher than Ca^{2+}, though Na is the most predominant cation. The mean $Mg^{2+}:Ca^{2+}$ ratio of these waters lies between 1 to 9, while a few samples have even higher $Mg^{2+}:Ca^{2+}$ ratio.

Szabolcs and Drab considered Mg per cent as calculated from equation (14) as one of the most important quantitative criteria for Mg hazards.

$$\text{Mg per cent} = \frac{Mg^{2+}}{\left(Ca^{2+}\right) + \left(Mg^{2+}\right)} \times 100 \tag{14}$$

A harmful effect on soil appears when the value exceeds 50 per cent.

The data of a pot experiment with wheat, using the normal sandy loam alluvial soil from Karnal and medium black soil (Vertisol) from Indore, India, indicated that the mean grain and dry matter yield of wheat decreased significantly with an increase in the $Mg^{2+}:Ca^{2+}$ ratio of the irrigation water, but the magnitude of decrease was greater at higher than at the lower electrolyte concentration (Table 6.10). It was further observed that the effect of these treatments was more pronounced in the heavy black clay soil than in the light-textured alluvial soil.

Table 6.10.   Effect of varying $Mg^{2+}:Ca^{2+}$ ratio of irrigation water on grain and dry matter yield of wheat in two types of soil and at different electrolyte concentrations

$Mg^{2+}:Ca^{2+}$ ratio in irrigation water	Grain yield, g/pot		Dry matter yield, g/pot	
	Type of soil			
	Alluvial soil	Vertisol	Alluvial soil	Vertisol
2	9.65	8.89	28.75	24.55
4	9.37	8.38	28.19	23.83
8	9.02	7.92	27.08	21.88
16	8.52	7.04	26.37	20.11
	Electrolyte concentration, me/l			
	20	80	20	80
	Alluvial soil		Vertisol	
2	12.97	5.57	33.94	19.36
4	12.77	4.98	33.51	18.56
8	12.46	4.48	31.70	17.26
16	12.07	3.50	31.27	15.25

Source: Gupta (1979).

It was found that $Mg^{2+}:Ca^{2+}$ ratio in exchange complex increased with an increase in $Mg^{2+}:Ca^{2+}$ ratio and decreased with an increase in SAR of the leaching water (Gupta, 1979). Furthermore, increasing the concentration of Mg^{2+} over Ca^{2+} in irrigation water increased the sodicity of the soil at a given SAR and electrolyte concentration of the water and its effect was more pronounced at higher than at lower SAR. Furthermore the release of Ca^{2+} from the soil and CaCO increased with an increase in $Mg^{2+}:Ca^{2+}$ ratio of the leaching water. With increase in $Mg^{2+}:Ca^{2+}$ ratio and SAR of the leaching water, the degree of soil dispersion increased significantly (Fig. 6.5). The

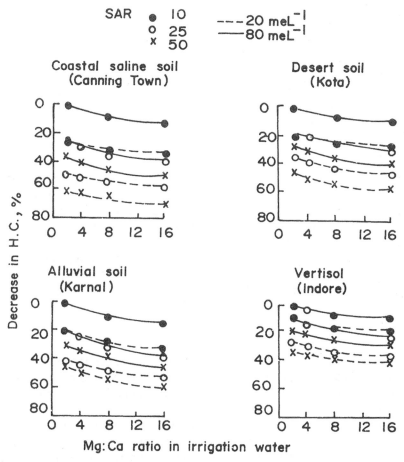

Fig. 6.5. Effect of varying Mg^{2+}:Ca^{2+} ratio on hydraulic conductivity of different soils (Gupta, 1979).

adverse effect of increasing Mg^{2+}:Ca^{2+} ratio was more pronounced at higher than at lower SAR and decreased with increase in the electrolyte concentration.

The role of Mg^{2+} vis-à-vis monovalent Na^+ and divalent Ca^{2+} in affecting soil proporties has been quite controversial. Because of its different affinity on clay complex, many scientists (Antipov-Karataev and Manaeva, 1958; Arnay, 1956) believe that it behaves more like Na^+ than Ca^{2+} ion. However, Quirk and Schofield (1955) and El-Swaify *et al.* (1970) did not find any difference between the behaviour of Ca^{2+} and Mg^{2+}.

Emerson and Chi (1977) observed that the ESP required to produce a given decrease in the flow rate was reduced when Mg^{2+} rather than Ca^{2+} was the complementary cation. Similarly, Yadav and Girdhar (1981) reported increased dispersion of clay particles and reduced hydraulic conductivity when Mg^{2+}:Ca^{2+} ratio was increased at a given SAR and electrolyte concentration of the leaching water. Rowell and Shainberg (1979) and Alperovitch *et al.* (1981) have reported that though Mg^{2+} decreased hydraulic conductivity of non-calcareous soils, it did not show any effect in calcareous soils. It appears that further research is required to understand the exact role of Mg^{2+} vis-à-vis Ca^{2+} on soil properties.

h) Nitrates

In certain parts of the world, the underground waters contain high concentration of nitrates. Normally, these waters also contain high concentration of potassium. Though high amounts of NO_3^- act as a source of N, yet continuous application of such waters results in higher vegetative growth and poor grain formation. The grains remain shrivelled and under developed. Thus, such waters are harmful for grain crops such like wheat, corn, barley, and gram, but can be successfully exploited to grow fodder crops. However, higher concentration of N in the fodder, as a result of high NO_3^- content of irrigation water, can cause NO_3^- –poisoning in animals.

i) Silica

High concentration of soluble Si in the irrigation waters can lead to hard crust formation similar to that of high RSC waters and thus prevent seedling emergence. So the waters containing high concentration of Si are equally unsuitable for irrigation.

Some other elements, including F, Mo, and Al, can be equally toxic for plant growth and/or deteriorate soil physicochemical properties. General guidelines relating to maximum permissible levels for these elements are given in Table 6.11.

6.4. Quality of Irrigation Water as a Function of its Origin

6.4.1. Rain Water

Rain water has the lowest salt content of all types of waters used for irrigation. This water contains dissolved gases (N_2, Ar, O_2, CO_2) and dissolved salts originating from terrestrial and marine sources. Generally, the amount of ions in rain water (NH_4^+, Cl^-, Na^+) varies widely and is dependent on the distance from the sea and the areas of aeolic deflation (2.20 t/km^2). In some polluted areas it may contain high concentration of SO_4^{2-} and its pH may be quite low, as in "acid rains" observed in many industrialised countries. This may cause serious leaf burn and damage the productivity of the soils.

Table 6.11. Recommended maximum concentrations, mg/l, of trace elements of irrigation waters[1]

Elements	For waters used continuously on all soils	For use up to 20 years on fine-textured soils at pH 6.0 to 8.5
Aluminium	5.00	20.00
Arsenic	0.10	2.00
Beryllium	0.10	0.50
Boron	0.75	0.2–10.0
Cadmium	0.01	0.05
Chromium	0.10	1.00
Cobalt	0.05	5.00
Copper	0.20	5.00
Fluorine	1.00	15.00
Iron	5.00	20.00
Lead	5.00	10.00
Lithium	2.50	2.50[2]
Manganese	0.20	10.00
Molybdenum	0.01	0.05[3]
Nickel	0.20	2.00
Selenium	0.02	0.02
Vanadium	0.10	1.00
Zinc	2.00	10.00

[1] These levels normally have no adverse effect on plants or soils.
[2] Recommended maximum concentration for citrus is 0.75 mg/l.
[3] Only for fine-textured acid soils,, or acid soils with relatively high content of iron oxide.
Source: Ayers and Westcot (1976).

6.4.2. Surface Water

The salt content of surface water is a function of the kind of rocks prevalent in the water's course, the nature of the soil over which the water flows, and the eventual pollution by human activities. Surface waters can be classified into flowing waters (rivers and streams) and stagnant waters (e.g., ponds and lakes). From the composition of the river waters throughout the world (Table 6.12), it is seen that the predominant anions are HCO_3^- and SO_4^{2-}, and the main cations are Ca^{2+} and Na^+.

Table 6.12. Ionic composition, mg/l, of river waters of the world

Source	Ion concentration							
	HCO_3^-	SO_4^{2-}	Cl^-	NO_3^-	Ca^{2+}	Mg^{2+}	Na^+	K^+
North America	68.0	20.0	8.0	1.0	21.0	5.0	9.0	1.4
South America	31.0	4.8	4.9	0.7	7.2	1.5	4.0	2.0
Europe	95.0	24.0	6.9	3.7	31.1	5.6	5.5	1.7
Asia	79.0	8.4	8.7	0.7	18.4	5.6	9.3	
Africa	42.0	13.5	12.1	0.8	12.5	3.8	11.0	
Australia	31.6	2.6	10.0	0.05	3.9	2.7	2.9	1.4

6.4.3. Underground Water

The salt content of underground water is dependent on the original source of the water and on the course over which it flows. Mineralisation of groundwater is in accordance with the law of dissolution, based on the contact between the water and the salt-bearing strata. Changes in the salt content of groundwater in the recharge process result from base exchange, transpiration, evaporation, and precipitation. In general, the salt content increase is due to evapotranspiration or dissolution, and is mostly affected by climate.

6.4.4. Sea Water

Sea water is a complex solution containing a large number of elements (e.g., ions, gases, organic matter, micro-fauna, and flora). Chloride (55%), sodium (30%), sulphur as sulphate (7%), magnesium (3.7%), and potassium (1.1%) are among the dominant elements present in the sea water. Sea water can be used for irrigation purposes only after undergoing an industrial desalinisation process and/or on highly drained soils in conjugation with good quality waters.

6.5. Water Quality Classifications

A number of classification systems of irrigation waters have been proposed which, in general, are based on the assumption of average conditions of good management (amount applied, drainage, texture, climate, crop tolerance, etc.). Any deviation from the average for these variables may require deviations from the general guidelines. Hence, a good water under average conditions may be marginal in some cases, while a marginal quality water may be used in other cases.

A widely used classification system is the one proposed by the US Salinity Laboratory Staff (Richards, 1954), to which several modifications depending on the local conditions have been added. It must be emphasized that this classification was proposed at a time (although it is based on extensive practical experience) when the quantitative aspects of the bicarbonate effects on the sodium hazards were but poorly understood. Therefore, the general guidelines based on this classification system must be regarded with some caution when critical levels of HCO_3^- are present and may have to be amended in the light of the contents of section 6.2.2. The US classification system is summarised in Fig. 6.6 and is based on C (EC, μ S/m) and S (SAR) of the irrigation waters, using the following equations:

Upper curve: $S = 43.75 - 8.87 \, Log \, C$ (15)

Middle curve: $S = 31.31 - 6.66 \, Log \, C$ (16)

Lower curve: $S = 18.87 - 4.44 \, Log \, C$ (17)

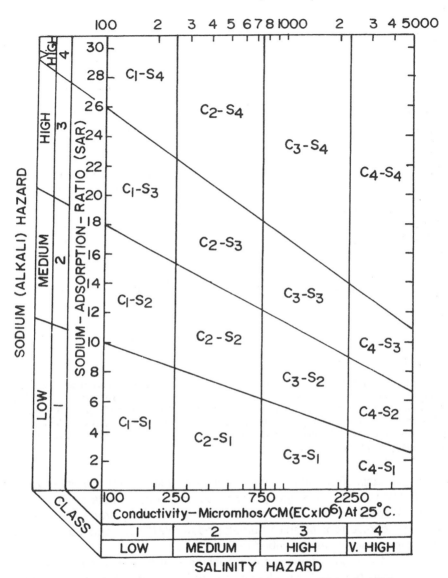

Fig. 6.6.   US Salinity staff diagram for classifying irrigation waters (Richards, 1954).

The "S-C surface" is, as a result, divided in 16 zones corresponding to all possible S-C combinations. The accompanying guidelines for use of this diagram are found in Tables 6.13 and 6.14.

Kanwar and Kanwar (1969) reported that, keeping in view the soil texture, the following water classes can be used for irrigating various types of crops (Table 6.15).

Table 6.13.   Guidelines for interpreting water quality for irrigation

Irrigation water quality index	Degree of problem		
	None	Increasing	Severe
Salinity (water availability) EC_{IW}, dS/m	< .75	0.75–3	> 3
Permeability (infiltration rate)			
EC_{IW}, dS/m	> .5	05–0.2	< 0.2
SAR_{adj}			
Montmorillonite	< 6	6–9[1]	> 9
Illite-vermiculite	< 8	8–16[1]	> 16
Kaolinite-sesquioxide	< 16	16–24[1]	> 24
Specific ion toxicity (affects sensitive crops)			
Sodium[2], (adj SAR)	< 3	3–9	> 9
Chloride[2], me/l	< 4	4–10	> 10
Boron, me/l	< 0.75	0.75–2	> 2
Miscellaneous			
$NO_3^- -N$ or $NH_4^+ -N$, mg/l	< 5	5–30	> 30
HCO_3^-, me/l (overhead sprinking)	< 1.5	1.5–8.5	> 8.5
Cl^-, me/l	< 1	1–2	> 2
pH	Normal range	6.5–8.4	

Source: Richards (1954).

[1]Values referring to dominant type of clay minerals. Problems are less likely to develop if water salinity is high and more likely to develop if water salinity is low. Use lower range if EC_{IW} < 0.4 dS/m, intermediate range if EC_{IW} is 0.4–1.6 dS/m, and upper limit if EC_{IW} < 1.6 dS/m.
[2]Most field crops and woody ornamentals are sensitive to Na^+ and Cl^-. With sprinkler irrigation on sensitive crops, Na^+ and Cl^- in excess of 3 me/l under certain conditions results in excessive leaf absorption and crop damage.

On the basis of the major components, i.e., salinity and sodicity hazards, the irrigation water can be grouped into four catagories (Table 6.16).

Considering the monsoon type of rainfall, choice of crops, texture and drainage conditions of the soil, Bhumbla and Abrol (1972) proposed guidelines for Indian conditions as described in Table 6.17.

Haque (1989), while taking into account the total dissolved solids, SAR, and RSC of the waters, and considering the degree of management and texture of the soil, proposed safe limits (Table 6.18) for the use of groundwaters for irrigation purposes.

6.6. Factors Affecting Suitability of Water for Irrigation

The following factors must be taken into account for considering the suitability of a given water for irrigation purposes:

Table 6.14. Use of irrigation waters according to the US Salinity laboratory Staff (1954): water quality classification

Salinity classification	Sodium classification
Low-salinity water (C1) can be used for irrigation with most crops on most soils with little likelihood that soil salinity will develop. Some leaching is required, but this occurs under normal irrigation practices except in soils of extremely low permeability.	*Low-sodium water* (S1) can be used for irrigation on almost all soils with little danger of the development of harmful levels of exchangeable sodium. However, sodium-sensitive crops such as stone-fruit trees and avocado may accumulate injurious concentrations of sodium.
Medium-salinity water (C2) can be used if a moderate amount of leaching occurs. Plants with moderate salt tolerance can be grown in most cases without special practices for salinity control.	*Medium-sodium water* (S2) will present an appreciable sodium hazard in fine-textured soils having high CEC, especially under low leaching conditions, unless gypsum is present in the soil. This water may be used on coarse-textured or organic soils with good permeability.
High-salinity water (C3) cannot be used on soils with restricted drainage. Even with adequate drainage, special management for salinity control may be required and plants with good salt tolerance should be selected.	*High-sodium water* (S3) may produce harmful levels of exchangeable sodium in most soils and will require special soil management— good drainage, high leaching, and organic matter additions. Gypsiferous soils may not develop harmful levels of exchangeable sodium from such waters. Chemical amendments may be required for replacement of exchangeable sodium, except that use of amendments may not be feasible with waters of very high salinity.
Very high-salinity water (C4) is not suitable for irrigation under ordinary conditions, but may be used occasionally under very special circumstances. The soils must be permeable, drainage must be adequate, irrigation water must be applied in excess to provide considerable leaching, and very salt-tolerant crops should be selected.	*Very high-sodium water* (S4) is generally unsatisfactory for irrigation purposes except at low and perhaps medium salinity, where the solution of calcium from the soil or use of gypsum or other amendments may make the use of these waters feasible.

a) chemical composition of irrigation water,
b) crops to be irrigated,
c) soils to be irrigated,
d) climate, and
e) management of irrigation and drainage.

The interaction of these factors can modify the limits otherwise imposed by the composition of irrigation water for their successful exploitation.

a) Chemical composition of irrigation water
The quality of the water is determined by the total salt content and by its ionic composition. The total salt content (expressed in g/l, me/l, TDS, or ppm) or the

Table 6.15.   Water quality class and its suitability for crops grown in soils of different textures

| Soil texture | Water quality class suitable for | | |
	Sensitive crops	Semi-tolerant crops	Tolerant crops
Clay	C_2S_1	C_3S_1	C_4S_1
		C_1S_2	C_2S_2
Loam	C_2S_1	C_3S_1	C_4S_1
	C_2S_1	C_2S_2	C_3S_2
		C_1S_3	C_2S_3
Sandy loam	C_2S_1		
	C_2S_2	C_4S_1	C_4S_1
	C_1S_3	C_3S_2	C_4S_2
		C_2S_3	C_3S_3
Loamy sand	C_3S_1	C_4S_1	C_4S_1
	C_2S_2	C_3S_2	C_4S_2
	C_1S_3	C_2S_3	C_3S_3
		C_1S_4	C_2S_4

Source: Kanwar and Kanwar (1969).

Table 6.16.   General classification of irrigation waters

Water quality rating	EC, dS/m	SAR	RSC, me/l
Good quality water	< 2	< 10	nil
Marginal water	2–4	< 10	< 2.5
Saline water	> 4	< 10	nil
Sodic water	< 4	> 10	usually > 2.5

electrical conductivity (dS/m) may give a general indication of the water's quality.

Also important is the determination of the main cations and anions, usually expressed in me/l, as several ion ratios influence water suitability.

b) Crops to be irrigated

In this chapter we are mainly interested in classifying water for agricultural purposes. Therefore, the crop is the first and most important factor to be considered. This evaluation must be based on the tolerance of a specific crop or crops in the rotation to the total salt content and or specific ion concentration in the irrigation water. The tolerance of a crop to salinity is that concentration of the soil solution that will give a minimum reduction in the yield as compared to non-saline conditions. Some crops are specifically sensitive to chloride and sodium ion concentration. The most important of these crops are the deciduous trees, citrus, and avocado.

Table 6.17.   Guidelines for interpretation of water quality for irrigation under Indian conditions

Soil type and texture	Crops to be grown	Upper permissible limit of EC of water for safe use for irrigation, dS/m
Deep black soils and alluvial soils having a clay content of more than 30%. Soils that are fairly to moderately well drained.	Semi-tolerant Tolerant	1.5 2
Heavy textured soils having a clay content of 20–30%. Soils that are well drained internally and have a good surface drainage system.	Semi-tolerant Tolerant	2 4
Medium-texutred soils having a clay content of 10–20%. Soils that are well drained internally and have a good surface drainage system.	Semi-tolerant Tolerant	4 6
Light-texutred soils having a clay content of less than 10%. Soils that have excellent internal and surface drainage.	Semi-tolerant Tolerant	6 8

Source: Bhumbla and Abrol (1972).

Qualifying remarks:
1. A monsoon rainfall of 300 to 400 mm is common for most areas having a groundwater quality problem. This rainfall periodically leaches out salts accumulated in the rootzone during the previous season.
2. In the above proposed limits of water quality it is presumed that the groundwater table at no time of the year is within 1.5 m from the surface. If the water table does come up within the rootzone, the above limits need to be reduced to half the proposed values.
3. If the soils have impeded internal drainage on account of presence of hard pans, unusually high amounts of clay, or other morphological reasons, the limit of water quality should again be reduced to half.
4. If the waters contain sodium greater than 70 per cent of total cations, gypsum should be added to soil occasionally.
5. If supplemental canal irrigation is available, water of higher electrical conductivity can be used in periods of water shortage.

Table 6.18.   Criteria for use of groundwaters for irrigation

Soil Texture	Other conditions	Maximum limits for safe use		
		TDS, mg/l	SAR	RSC, me/l
Coarse	Level lands, good water management	1500	12	5.0
Medium	Fairly level lands, adequate water supply	800	10	2.5
Fine	Level lands, poor water supply, low level of management	500	5	1.25

Source: Haque (1989).

c) Soils to be irrigated

The behaviour of a soil in contact with saline water depends on its initial physical properties and on the salt content of the water. The nature and quantity of clay in the soil affects the ion adsorption capacity, which in turn influences the hydrophysical properties. Furthermore, the presence of an impermeable layer or of a groundwater table affects the salt distribution in the profile.

The application of saline water to a salt-free soil will salinise the soil, but the use of the same quality of water may reduce the salinity of a saline soil if drainage is adequate. As infiltration and percolation of water may differ greatly for different soils, different degrees of salinisation may be expected with the same quantity and quality of irrigation water.

d) Climate

Evapotranspiration and rainfall are the two main climatic elements to be considered when evaluating suitability of water for irrigation. The water depth to be applied to a crop during a season depends on the evapotranspiration, which therefore affects the irrigation regime and consequently the seasonal dynamics of salts in the soil profile. In general, irrigation water of relatively high EC can be used in areas of low ET demand.

e) Management of irrigation and drainage

The irrigation method influences the salt accumulation in the soil and the capacity of the plant to tolerate the developed salinity. The application of amounts of water less than the consumptive use will result in accumulation of salts in the rootzone, while provision of drainage will enable the use of even medium to hazardous irrigation waters.

7

Wastewaters as a Source of Irrigation

Exponentially increasing urbanisation is not only using large areas of productive agricultural land but is also producing large volumes of wastewaters (sewage and other effluents) that have become a serious environmental threat in most countries. These waters are a direct source of surface waterlogging, groundwater contamination, and salinisation of good quality lands around cities.

Not only degradation of good quality lands but also many water pollution problems have been created by this uncontrolled disposal of untreated wastewaters. As a result, pathogens causing cholera, hepatitis, and malaria are increasing and becoming a threat to human and biotic life. Also, high amounts of toxic elements, oxygen demanding organic wastes, dissolved salts and other chemicals, and suspended particles in these wastewaters are causing serious damage to our environment.

Stagnating waters are a cause of foul smell and serve as a fertile breeding ground for mosquitoes, because of which incidents of malaria and fillaria are increasing every day. One significant way to eradicate malaria is to eliminate such breeding grounds. However, at present, chemicals such as DDT, BHC, and melathion are being sprayed. In India alone, 1,000 tonnes of DDT is used every year for this purpose. Instead of solving the problem, this further pollutes the environment with deadly insecticides that are known to have long residual life and can pass through the food chain even to mother's milk.

Mostly, the municipalities collect sewage through the sewer system and pump it out directly from the main sewer or collection tank into a nearby drain, stream or river body that ultimately joins the major river of the area. Most industries discharge their effluents into rivers or canals causing serious deterioration in the quality of their waters. As a result, almost all the rivers including the Ganga and Yamuna, and many sea beaches have been polluted to an alarming level in India. This is also the case in many developing countries. Where such outlets are not naturally available, sewage is simply dumped into a nearby low-lying area or pond. As a result of such an unplanned disposal of municipal wastes, many big lakes of sewage can be seen in and around all big cities. In fact, all the rivers have become synonymous with big open sewer lines and lakes and ponds with sewage dumps.

Of the large varieties of wastewaters, sewage, which is available throughout the year in the vicinity of the urban communities, is a potential source of irrigation and nutrients for biomass production. In those parts of the world where water, due to low rainfall and poor local supplies, is insufficient and fresh water is brought from outside (through canals), wastewater reuse becomes still more important. In other areas, conserving wastewaters can delay and/or reduce the depletion of local sources of natural waters.

7.1. Quantity of Wastewaters

Based on the population, community water supply, and method and efficiency of collection, the amount of domestic wastewaters produced may vary from place to place. It is estimated that the production of sewage water (SW) alone may vary from 12,500 (Chhabra, 1988 a) to 3,650 million l/day (Juwarkar, 1988) in India. In general, it varies from 50 to 120 l/day per person.

The wastewater produced by an industrial unit will depend upon the type and size of industry, the process being used, and the extent of its reuse within the industry. However, it is estimated that around 7,500 million l/day (MLD) of wastewater is generated by industrial units in India.

7.2. Quality of Wastewaters

The quality of industrial wastewaters (effluents) varies from industry to industry and depends upon the raw material, end-product, process of manufacturing, extent of its reuse within the industry, and quality of raw water. The quality of sewage differs from place to place and depends upon season, per capita consumption, and quality of the community water supply.

7.3. Irrigation and Nutrient Potential of Wastewaters

Domestic and industrial wastewaters of suitable quality can be used directly or after some treatment, to meet the ever-increasing demand of water and nutrients and thus help to maximise production of food, fibre, and fuel (Duston and Lonsford, 1955; Fisk, 1964; Sachin and Menon, 1976; Juwarkar, 1988). Using Israelson and Hansen's (1967) equation $a = q \cdot t/d.36$, which gives the relationship between area in hectare (a), quantity of water available in m^3/sec. (q), time in hours required for irrigation (t), and the depth of irrigation in cm (d), one can estimate the potential of the available wastewater for irrigating different crops. Chhabra (1988a) reported that in India alone, SW has the potential to irrigate 250,000, 380,000 and 330,000 ha of vegetables, fodders and cereals, respectively (Table 7.1). Considering that a shallow cavity tubewell can irrigate 4 ha of land (Tyagi *et al.*, 1980), the irrigation potential of SW alone is equal to that generated by 65,000 to 95,000 tubewells. Similar estimates can be made for other countries.

Table 7.1. Irrigation potential of sewage waters of India

Type of crop	Depth of irrigation, cm	Irrigation interval, days	No. of irrigations per crop	Irrigation potential, in 1,000 ha
Vegetables	5.0	10	10	250
Fodders	5.0	15	10	380
Cereals	7.5	20	5	330

Source: Chhabra (1988 a).

The nutrient contents of wastewaters vary greatly and depend upon their origin. In the industrial effluents, concentration of P and K is much lower (except in organophosphate- and P-fertiliser-based industries), while that of N and S is generally moderate to high.

In general, SW are rich in N, P, K, S, and micronutrients (Table 7.2). Based on the average values, it is estimated that through five irrigations, each of 7.5 cm, untreated SW can add 181, 29, 270, and 130 kg/ha of N, P, K, and S respectively, which is enough to meet the nutrient requirement of the crop. In addition, 1.28, 0.75, 41.86 and 1.37 kg of Zn, Cu, Fe, and Mn respectively, are added to meet the micronutrient requirements. It is further estimated that SW in India have a potential to contribute 743,370 t of nutrients per year (Table 7.3).

Table 7.2. Nutrient concentration in the sewage waters of Haryana, India

Nutrient	Conc., mg/l	
	Range	Mean
Nitrogen	25.38– 97.58	48.26
Phosphorus	4.34– 12.71	7.58
Potassium	27.69–152.10	72.44
Sulphur	19.02– 60.93	34.59
Zinc	0.13– 0.90	0.34
Iron	0.59– 21.81	10.83
Copper	0.06– 0.61	0.20
Manganese	0.25– 0.60	0.36

Source: Baddesha *et al.* (1986).

Table 7.3.   Nutrient potential of sewage waters of India

Nutrient	Average conc., mg/l	Contribution potential		Price of nutrients*** US$	Economic value, US$, Million
		t/day	t/year		
Nitrogen (N)	48 ± 14 **	600	219,000	260	56.94
Phosphorus (P)	8 ± 3	100	36,500	670	24.46
Potassium (K)	72 ± 9	900	328,000	119	39.03
Sulphur (S)	35 ± 3	438	159,870	106	16.95
Total					137.38

* Average flow of SW, 12,500 million l/day.
** SE.
*** Based on prices for 1983.
Source: Chhabra (1988a)

Thus, use of wastewaters saves expenditure on irrigation as well as on plant nutrients. Furthermore, as these nutrients are provided with each irrigation, there is no additional cost of their application. Their efficiency of utilisation is also high, as is the case in split application of nutrients.

7.4. Pollution Hazards

When wastewaters are discharged into water bodies (rivers, streams, sea, drains, or lakes), they can be a serious source of pollution beacuse of their high biological oxygen demand (BOD), chemical oxygen demand (COD), suspended solids, high concentration of nutrients, toxic elements and pathogens. On the basis of their composition, these waters have been rated as strong, medium, and weak pollutants (Table 7.4). In view of the BOD, free ammonia, and total suspended solids (Table 7.5), most of the untreated

Table 7.4.   Characteristics of untreated sewage affecting the pollution of river bodies

Characteristic	Maximum concentration, mg/l for pollution rating		
	Strong	Medium	Weak
Total solids	1,200	720	350
Suspended solids	350	220	100
Settleable solids	20	10	5
BOD_5	400	220	110
OC.	290	160	80
COD	1,000	500	250
Total N	85	40	20
Free ammonia	50	25	12
Total P	15	8	4
Cl⁻	100	50	30
Grease	150	100	50

Table 7.5. Characteristics of untreated sewage waters of Haryana, India

Characteristic	Range	Mean
Pollution parameter		
pH	7.00– 7.50	7.20
BOD_5, mg/l	66.00–173.00	125.00
TS, mg/l	0.57– 2.23	1.44
Free ammonia, mg/l	9.80– 32.20	17.90
OC, mg/l	93.00–243.00	176.00
Irrigation quality parameter		
EC, dS/m	0.93– 2.87	1.68
SAR,	1.56– 5.96	3.17
RSC, me/l	nil	nil
Na^+, me/l	2.97– 18.42	7.97
Ca^{2+}, me/l	2.80– 5.67	3.84
Mg^{2+}, me/l	4.02– 15.33	8.62
CO_3^{2-}, me/l	absent	absent
HCO_3^-, me/l	4.30– 11.42	7.22
Cl^-, me/l	1.71– 12.35	6.00
SO_4^{2-}, me/l	1.19– 3.81	2.14

Source: Baddesha *et al.* (1986).

sewage waters are unsafe for disposal in rivers or other water bodies. However, for land disposal, these waters are considered weak to medium (Metcalf and Eddy, 1972) in their pollution hazards.

Total solid contents of the wastewaters are normally high and may cause siltation of water bodies when discharged without any pretreatment. However, when disposed of on land, they do not create any problem.

7.5. Quality Criteria for Reuse of Wastewaters for Irrigation

The norms for wastewater utilisation through irrigation need to be examined for irrigation quality as well as public health (both toxic ions and pathogens) and are discussed below.

7.5.1. Salinity, Sodicity and Specific Ion Hazards

The criteria to judge the suitability of wastewates reuse with respect to salinity, sodicity, and specific ion hazards are the same as for the normal irrigation waters and have been discussed in the previous chapter. Normally, the effluents from chemical industries (Table 7.6) have unfavourable pH, either too low or too high, affecting metal ion solubility and soil acidity or alkalinity. These effluents also contain high amounts of Na^+, SO_4^{2-}, NO_3^-, and Cl^- which

Table 7.6.   Characteristics of wastewaters from organophosphorus pesticide factory

Characteristic*	Source of wastewater		
	Phosalone	Melathion	Diazinon
pH	9.4	11.5	9.2
COD	30,343	43,893	66,080
BOD$_5$	10,980	21,150	23,570
Total solids	59,000	39,920	37,148
Total dissolved solids	57,040	39,620	
Suspended solids	1,960	300	700
P (inorganic)	510	17	
P (organic)	2,490	4,182	270
SO$_4^{2-}$	1,120	2,600	
Cl$^-$	42,738	5,440	61,875
Na$^+$	23,600	12,400	
K$^+$	120	360	

*All values except pH are expressed in mg/l

restrict their use for agricultural production. Such waters need treatment within the industry before their reuse.

The effluents from food industries normally contain less amount of salts and thus pose a low risk of creating salinity or sodicity problems in soil. Though these effluents contain high amounts of OC and thus have high BOD yet when used in a proper manner they do not cause oxygen stress to most agricultural crops. Furthermore, as these effluents contain high amounts of nutrients (Table 7.7) they can be used as liquid fertilisers to boost crop yields.

Table 7.7.   Nutrient and organic carbon content, mg/l, of industrial wastewaters suitable for crop irrigation

Wastewater source	N	P	K	OC
Fruit and vegetable canning	16.0–30.0	0.5– 5.3	10.0–107.1	580.0–696.0
Sugar refining	30.0–60.0	1.3– 1.8	91.3–116.2	2900.2–3480.3
Starch	43.0–99.0	0.5–0.8	13.3–16.6	1160.1–1740.1
Hemp	37.0–90.0	2.6–16.3	303.8–727.1	928.1–2540.6
Textile mill	30.0–35.0	6.6–8.8	8.3–12.5	
Dairy	40.0–50.0	2.9–5.3	8.3–10.0	464.0–580.0
Mango processing	15.0–20.0	1.8–5.3	8.3–12.5	580.0–1450.0
Tomato processing	10.0–15.0	2.2– 3.5	6.6–12.5	464.0–870.0

Source: Adapted from Juwarkar (1988).

In general, sewage waters are neutral to slightly alkaline in reaction and low in salt content (Table 7.5). Magnesium is the most dominant cation in SW, followed by Na^+ and Ca^{2+}. Though municipal waters of only a few places contain more Mg^{2+} than calcium, almost all the SW contain more Mg^{2+} than calcium. However, as these waters have Mg^{2+} : Ca^{2+} ratio less than 4 and SAR less than 10, they are not likely to pose Mg^{2+} hazards.

Among the anions, CO_3^{2-} are absent, while HCO_3^- are more numerous than Cl^- and SO_4^{2-}. However, RSC is nil in these waters. Based on quality rating, soil and agroclimatic conditions, waters of EC, SAR, and RSC less than 2 dS/m, 10 and 2.5 me/l respectively are of 'A' quality (Manchanda, 1976) and can be used for irrigation. Most of these waters fall in C_1S_1 and C_2S_1 classes as per the criteria of USDA (Richards, 1954). Normally, high concentration of Cl^- in SW is due to human excrement and not to chlorination of municipal waters.

7.5.2. Contamination Due to Pathogens

The presence of enteric pathogens in wastewaters (Table 7.8) is the most serious hazard. They are capable of causing many diseases and disorders directly through contact to farmworkers and animals and indirectly through

Table 7.8. Pathogens in the wastewaters and diseases caused by them

Pathogen	Disease
Bacteria	
Escherichia coli	Diarrhoea
Salmonella typhi	Typhoid fever
Salmonella paratyphi	Paratyphoid fever
Other *Salmonella* spp.	Foodpoisoning Salmonellosis
Shigella spp.	Bacillary dysentery
Vibro cholerae	Cholera
Other *Vibro* spp.	Diarrhoea
Clostridum perfringens	Gastroenteritis
Yersinia enterocolitis	Diarrhoea and septicaemia
Bacillus cereus	Gastrocenteritis
Virus	
Polio virus	Poliomycetis
Hepatitis A virus	Hepatitis
Coxsackie virus	Coxsackie infection
Rota viruses	Diarrhoea
Helminths	
Ascaris lumbricoides	Ascariasis
Trichuris trichicaura	Trichiriasis
Ancyclostome duodenals	Ancyclostomiasis
Schistosoma mansoni	Schistosomiasis
Protozoa	
Entamoeba histolytica	Amoebiasis Colonic ulceration
Giardia lamblia	Giardiasis

contamination of soil and consumption of contaminated crop, vegetable, and fodder, to the consumer. These organisms can survive in soil from 6 to more than 100 days, depending upon soil conditions (Table 7.9). Neutral to slightly alkaline pH, medium to heavy texture, and conditions leading to excessive

Table 7.9.   Survival of enteric pathogens in the soil and on the crops*

Pathogen	Type	Survival in days
Survival in the soil		
Salmonella typhi	Peat	13
	Sandy soil	29–36
	Muck	12
	Clay loam	12–20
	Loam	21–120
Salmonella spp.	Pasture lands	200
	Sandy soil	35–84
Streptococcus faecallis	Sandy loam	28- 77
	Loam	26–63
	Clay loam	33–39
	Muck	40–77
Entamoeba histolytica	Moist soil	6–8
Ascaris ova	Irrigated soil	730–1035
Survival on the crop		
Salmonella spp.	Fodder	12–42
	Root crops	10–53
	Leafy vegetables	1–40
	Radish	28–53
	Lettuce	18–21
	Carrot	10–15
	Tomato	3–20
Shigella spp.	Fodder	2–5
	Vegetables	2–7
	Orchard crops	6–10
	Tomato (surface)	2–3
	Tomato (tissue)	10–12
Vibro cholerae	Vegetables	5–7
	Lettuce	5–10
Ascaris ova	Lettuce tomato and leafy vegetables	2–5
Entamoeba histolytica	Leafy vegetables	15–60
Entero virus	Vegetables	15–60

*Compiled from different sources

moisture help them to survival in soil. Though, as a result of land application of wastewaters, their active population goes down they can form cysts, and hibernate for a long time and become active when the environment is suitable for their growth and multiplication. On crop, these pathogens can survive up to 60 days (Table 7.9). Though no bacterial penetration is possible through healthy surface of fruits and vegetables, their outer surface may get contaminated with pathogens directly through irrigation and indirectly via contact with the polluted soil. Once they are contaminated, even four or five washings are not sufficient to make the vegetables free from pathogens. Kruse (1962) reported that even heavy rainfall did not wash away *Coliform* from clover that was irrigated with settled sewage. Broken, brushed, or damaged portion of the fruits and vegetables can easily allow the pathogens to enter into it and hibernate for a long time. Therefore, wastewater must be treated so as to disinfect it from pathogens (a very costly process) or crops selected that are not directly consumed. It is strongly suggested that crops used raw (uncooked) and whose edible parts are directly in contact with the wastewaters and/or soil should not be irrigated with contaminated wastewaters.

7.5.3. Toxic Elements

The concentration of toxic elements (Zn, Cu, Fe, Mn, Pb, Cd, Ni, Hg, As, Cr, Se etc.) in the wastewaters is an important factor to be considered in the planning of land-based waste management. Though some of these are essential plant nutrients, almost all become phytotoxic at high levels. Concentration of toxic elements (Table 7.10) in untreated SW of many developing countries is generally within safe limits, because of the low level of industrialisation. However, in many cities where industrial effluent without any pretreatment is allowed to mix with the domestic disposal system, concentration of certain toxic elements may go above the critical limits. For this, efforts

Table 7.10. Trace element concentration in the untreated sewage of Haryana, India, and quality criteria for continuous irrigation use

Element	Av. conc., mg/l	Irrigation quality criteria*	
		Coarse-textured soil	Fine-textured soil
Zn	0.34	2.00	10.00
Fe	10.83	5.00	-
Cu	0.20	0.20	5.00
Mn	0.36	0.20	-
Pb	0.09	5.00	20.00
Cd	0.01	0.01	0.05
Ni	0.14	0.20	2.00

*EPA (1973).
Source: Baddesha and Chhabra (1985)

should be made not to permit the disposal of industrial effluents into the domestic wastewaters. The industrial effluents should generally be given a specific treatment, depending upon the nature of the toxic metal, to minimise their pollution hazards.

7.6. Systems for Management of Municipal Wastewaters

Disposal of sewage is an expensive and sensitive problem. The municipal committees have administrative, legal, and moral responsibilities to dispose it in a manner that provides a clean environment to the public and makes this waste resource usable for the benefit of mankind. This can be achieved by one of the following systems:

7.6.1. Sewage Treatment Plants

Purification of sewage in a treatment plant involving primary, secondary and tertiary treatment is an ideal solution to this problem. By this process, sludge is removed from the sewage. This reduces its suspended particles, BOD, and concentration of toxic elements. The treated water can be reused for irrigation purposes and/or safely discharged into a natural water body. However, treatment plants are costly and out of reach for most municipal committees. In addition to the initial high cost, the recurring operational expenditure and the energy required to run these treatment plants is considerably high and is generally beyond the means of municipal committees.

Furthermore, with increasing industrialisation, more and more industrial effluent, particularly from small-scale industries such as galvanising plants, tool factories, tanneries, cardboard and handloom industries, which do not have any effluent treatment facility, is getting into the sewer system. This increases the load of pollutants in sludge, making it unfit for agricultural purposes. Already in Europe and many parts of the developed countries disposal of this contaminated sludge is posing a big problem. The treated water, though low in BOD and toxic elements, is yet not free from pathogens and soluble salts.

7.6.2. Oxidation Ponds and Fish Culture

In this system, sewage is stored for about 24 to 48 hrs in shallow but big pits known as oxidation ponds and after that allowed to flow into the nearby natural drain or river. Its BOD thus gets reduced and most suspended particles get settled. Treating sewage in oxidation ponds is less costly, especially when it is combined with pisciculture. However, it needs a large amount of space to store large volumes of sewage in shallow depths. The continuous presence of water generates foul smell and acts as a breeding ground for mosquitoes. Furthermore, there is a continuous downward seepage of sewage that may contaminate groundwater. Since certain categories of fishes can grow well in sewage ponds and utilise the nutrients and fauna developed there, pisciculture in oxidation ponds is very encouraging. However, fishes are known biological accumulators of toxic elements such as Pb, Cd, Ni, Zn, and Hg, which also decrease their own production and quality, especially flavour. Thus, fish production in untreated or partly treated sewage ultimately helps pass on the contamination to the consumers—human beings. Production of fish decreases when chemical effluents containing ammonia and organic compounds get mixed with the sewage. These are serious limitations to the use of this method.

7.6.3. Soaking Pits and Lagoons

Land disposal is being increasingly recognised as the cheapest and most cost-effective method of disposing sewage and other effluents. Soaking pits and lagoons are the most common forms of land disposal. In this system, sewage is allowed to go into the pit without any pretreatment, from where it is ultimately soaked into the soil. Under such a system, there exists a serious danger of soil and groundwater contamination. Also, these soaking pits get choked after some time and new pits must be made at another place. Such systems can serve well for an individual household, but are not practicable when large volumes of sewage are to be disposed of. In addition to the serious pollution problems, these techniques do not allow the use of wastewaters for agricultural production.

7.6.4. Sewage for Irrigated Agriculture

The composition of untreated sewage is generally suitable with regard to both pollution and irrigation water quality for its use in agriculture in most countries. Tapping this resource can save tremendously on fertiliser and irrigation costs and can help to boost agricultural production in arid and semi-arid countries. However, at present only 10 to 40 per cent of wastewater is being used for growing vegetables and that too only in the vicinity of big towns.

Use of sewage for raising crops, commonly known as sewage farming, seems to be favoured by municipalities. In many places municipal committees are even earning huge sums through auction of sewage. However, for optimum growth, plants should be supplied with water and nutrients as per their requirement, which varies with their physiological growth stages. Most of the plants do not require continuous application of nutrients, especially near the maturity stage. As sewage contains large quantities of nutrients, its continuous application increases soil fertility, which results in extra vegetative growth, delayed maturity, and lodging of the grain crop. Thus grain filling and harvesting index of the crop are adversely affected. The cost of harvesting also increases. Under such conditions farmers must have two independent sources of irrigation, one of the effluent and the other of good quality water, to ensure the success of the crop.

Irrigation with sewage makes the plant more succulent and thus more vulnerable to attacks by insects and pathogens. It also encourages the growth of weeds. To control these, farmers are forced to spray insecticides, fungicides, and weedicides, which further contaminate the vegetables. Carrot, turnip, potato, sweetpotato, and other vegetable with edible parts in direct contact with the soil are severely contaminated by the pathogens present in sewage. The raw consumption of radish and carrot produced at the sewage farms exposes people to serious health problems. Hence, a minimum degree of sewage pretreatment, appropriate crop selection (Table 7.11), and efficient water and soil management practices are key to successful sewage farming.

In view of some of the above disadvantages and other factors such as contamination due to toxic elements, organic compounds, pesticides, foul smell, and attitude of the consumers, sewage is not being fully utilised and should not be encouraged for agriculture in general and olericulture in particular.

Table 7.11. Guidelines for choice of vegetative cover in relation to degree of sewage treatment

Degree of treatment	Suggested vegetation, in order of preference
Untreated (raw) sewage	Forest and avenue trees, ornamental and flowering shrubs.
Primary treated sewage (preferably diluted or with alternate source of irrigation)	Industrial crops such as cotton, jute, milling-type sugarcane, sugarbeet, cigarette tobacco.
	Essential oil-bearing crops such as citronella, mentha, lemon grass.
	Crops raised exclusively for seed production.
	Cereals and pulse crops with well-protected grains such as wheat, rice, green gram, pigeonpea, black gram, soybean, sorghum, pearlmillet.
	Oilseed crops such as linseed, caster, *Brassica* spp., safflower, sunflower, sesamum.
	Well-protected fruit crops such as coconut, banana, citrus. Orchard crops where fruits are borne on the plant sufficiently far from soil, e.g., guava, grape, papaya, mango, sapota.
	Vegetables always cooked before eating and borne on the plant away from soil, e.g., lady's finger (okra), beans, brinjal (eggplant).
	Fodder crops such as maize, oats, teosinte, berseem, lucerne, shaftal.
Secondary treated sewage	All of the above types of vegetation. All types of crops including vegetables borne near the soil surface but consumed after cooking.
Tertiary treated and disinfected sewage	All crops without restriction. Pastures for milching cattle. Lawn grass and public gardens.

Source: Chhabra (unpublished).

7.7. Sewage Utilisation through Forestry

For safe utilisation of irrigation and nutrient potential of sewage, the following points must be considered (Chhabra, 1987b):

a) Sewage should not stagnate so as to produce foul smell and provide a breeding ground for mosquitoes.
b) It should not percolate down to contaminate groundwater and raise the water table. The disposed water should either evaporate or be consumed by the vegetation.
c) It should not increase the salinity of soil and/or make it unproductive by means of accumulation of toxic elements and deterioration in physical properties.
d) Any vegetation raised with sewage should not be directly consumed by human beings.
e) It should be an economical system generating employment and revenue.

In light of the above, a technology involving forestry has been developed at the Central Soil Salinity Research Institute, Karnal, India, wherein waterlogging, stagnation, and thus pollution are eliminated by utilising the irrigation and nutrient potential of these waters, right in the area of their production.

7.7.1. The Technology

The method consists of growing trees on ridges 1 m wide and 50 cm high (Fig. 7.1) and disposing of untreated sewage in furrows or shallow trenches (2 m wide). The amount of sewage or effluent to be disposed of in the system

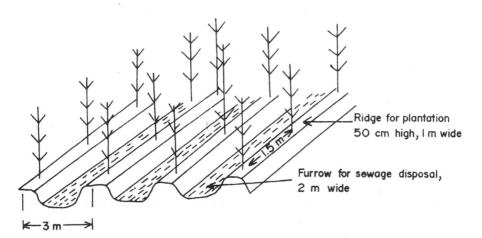

Fig. 7.1.   Schematic diagram showing ridge and furrow system for planting trees and disposal of raw sewage (Chhabra, 1989).

depends upon the type and age of the plant (canopy development), climatic conditions, soil texture, and quality of the untreated effluent. The total discharge of the effluent is so regulated that effluent disposed of is consumed within 12 to 18 hrs and there is no standing water left in the trenches. Through this technique, it is possible to dispose of 5 to 15 cm (0.3 to 1 million l) of the effluent per day per hectare and grow lush green trees.

The disposed water is consumed through the following processes:

a) Evaporation:

Depending upon the climatic conditions, the potential evaporation of most places varies from 0.1 to 1.0 cm/day (Fig. 7.2). In a compact forest area, as the tree canopy develops, it intercepts the radiant energy, lowers the ambient temperature, and through transpiration increases the relative humidity of the area. Both these factors reduce the potential evaporation. So the maximum amount of water that can be lost through this process is 1 cm/day', i.e., only 6.6 per cent of the total, if the disposal rate (loading) is 15 cm/day/ha.

b) Deep percolation losses:

The amount of water that can be lost to the groundwater through deep percolation will depend upon the permeability of the soil, which in turn depends upon the soil texture, structure, presence of impermeable layers (hard pan or $CaCO_3$ layer) in the soil profile, and the ESP of the soil. The nature of the soil selected should be such that it does not allow rapid infiltration (Table 7.12) of

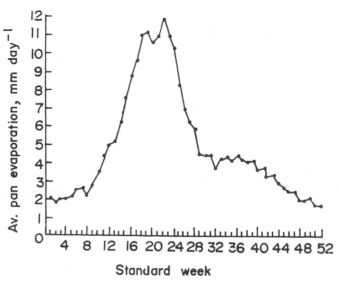

Fig. 7.2.   Average daily evaporation rate during different standard weeks at Karnal, India (Chhabra, 1989).

Table 7.12. Permeability of soil and its suitability for sewage disposal

Permeability class	Permeability rate, cm/day	Suitability class for sewage disposal
Very slow	< 3.60	Highly suitable
Slow	3.60–12.00	Suitable
Moderately slow	12.00–36.00	Suitable for intermittent disposal
Moderate	36.00–120.00	Not suitable
Rapid	> 120.00	Not suitable

the disposed water but retains it either on the surface and/or in the effective rootzone, from where it should be consumed through the process of evapotranspiration. Sandy soils, which have high permeability rate, may thus be unsuitable. But on such soils one can regulate the loading rates in such a way that one discharges smaller amounts but more frequently so as to minimise deep percolation losses and prevent rise in groundwater and its contamination.

c) Transpiration:
In this technology, trees are used as *biopumps,* absorbing water from the soil, using a part of it for their biomass production, and releasing the rest in the environment through transpiration. The amount of water lost through this process depends upon the amount of canopy developed (total transpiration area), the leaf physiology (the number of stomata per unit area), and the climatic factors. For normal cropped area, the evapotranspiration rates are equal to or less than the potential evaporation. However, transpiration in trees, because of their large canopy and the high transpiring area, can be many times greater than the potential evaporation. Chhabra (1988, unpublished data) from lysimeter studies observed that under situations where availability of water in the rootzone is not a limiting factor, 6- to 12-month-old eucalyptus plants can transpire 6 to 10 times more water than the potential evaporation of the area. It is generally observed that after rainfall, soils under eucalyptus plantation dry more quickly than under other vegetation. This is due to the fact that eucalyptus plants transpire about 90 per cent of water received through rainfall within a day or so. This capacity of eucalyptus to transpire huge quantities of water, which is otherwise not favourable for conserving rain water and sweet under-ground waters, has been used to biodrain marshy and swampy lands in Russia and Australia and is exploited in this technology to *biodrain* excess amounts of wastewaters.

In this technology the major emphasis is to select such tree species that can transpire huge quantities of water and that too throughout the year so that biodrainage of excess water is achieved at no extra cost. It is estimated that about 60 to 90 per cent of the discharged effluent can be biodrained through transpiration. As a consequence, the applied effluent normally disappears within 12 to 18 hrs resulting in no water stagnation, which in turn lowers the production of foul smell and eliminates breeding places for mosquitoes and other pathogens. Analysis of the ambient air from the disposal site has shown no extra methane production, confirming that such a system does not encourage anaerobic decomposition of organic materials contained in the untreated sewage and thus curtails the production of foul-smelling gases. In fact, this technology utilises the entire biosystem: soil as the dynamic living filter because of its physical, biological, and chemical properties for retention, degradation, absorption and precipitation of both organic and inorganic constituents of wastewaters, and the trees for utilisation of nutrients and water through assimilation for biomass production and to biodrain excess water through transpiration. It builds up the soil fertility (Fig. 7.3) with respect to soil available N, P, K, and OC and micronutrients (Fig. 7.4). It decreases the soil

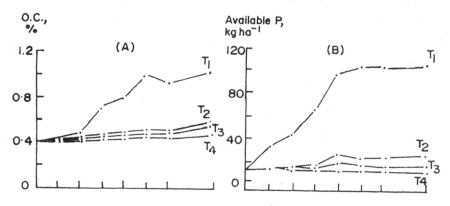

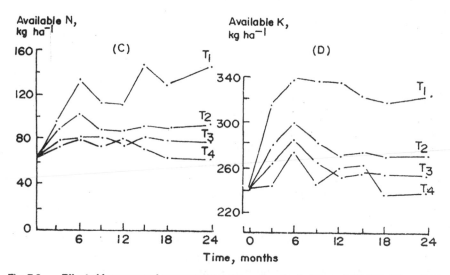

Fig. 7.3.　Effect of frequency of sewage water disposal on the build-up of (A) OC, (B) available P, (C) available N, and (D) available K in the surface 15 cm soil (Chhabra, 1989).

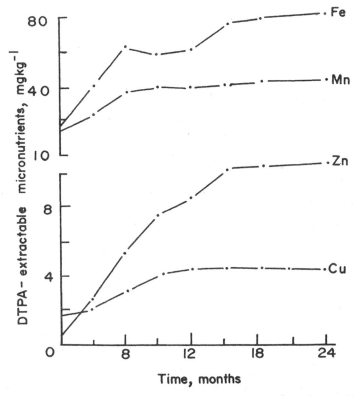

Fig. 7.4. DTPA-extractable micronutrients of the surface 15 cm soil as influenced by sewage water disposal every day (Chhabra, 1989).

pH (Fig. 7.5 a) from highly alkaline to neutral levels, without significantly building up the salinity (Fig. 7.5b) and thus ameliorates the highly deteriorated alkali soils. There is no adverse effect of sewage disposal due to toxicity or deficiency of any nutrient, and heavy metal or salinity stress on plants.

Since forest plants are used for fuel wood, timber, or pulp, there is no chance of pathogens, heavy metals or organic compounds entering the human food chain, which is a greater possibility when vegetables or other crops are raised with sewage.

7.7.2. Suitable Tree Species and their Response to Untreated Sewage

Though most tree species are suitable for utilising the effluent, yet those that are fast growing, can transpire huge amounts of water, and can withstand high moisture content in the root environment are most suitable. Eucalyptus (*Eucalyptus hybrid*; Fig. 7.6 a, b) has the capacity to transpire large amounts of water and remains active throughout the year. Other suitable species are

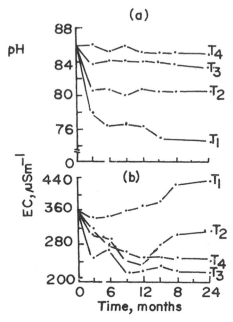

Fig. 7.5.   Changes in (a) pH and (b) EC of the surface soil as affected by sewage water disposal and eucalyptus plantation (Chhabra, 1989).

poplar (Fig. 7.7) and leucaena (*Leucaena leucocephala*). Out of these three species, poplar is most responsive in utilising sewage (Table 7.13, Fig. 7.8). However, being deciduous, it remains dormant and thus can not biodrain effluent during winter months. Eucalyptus, therefore, seems to be the best option when large volumes of effluents are to be disposed of on a limited area and throughout the year. Where large areas of wastelands are to be planted and the volume of effluent available is small, a combination of poplar and eucalyptus in blocks is the best proposition. At the same time, for such areas, instead of disposing sewage every day on the same piece of land, one can plant a larger area with the same volume of available effluent by actually using the effluent as a potential source of irrigation.

7.7.3. Type of Land Suitable for Such Plantation

As the sewage water itself provides nutrients and irrigation and ameliorates alkali soils by lowering pH, relatively infertile wastelands can be used for forest plantation. Areas along the roads, railway tracks, drains, old mines, and landfills after proper land shaping can be profitably exploited for this purpose. Such areas near the cities can be developed into scenic parks and mini woodlots. Undernourished, low density reserve forests can be rehabilitated; government and community wastelands can be brought under productive

Fig. 7.6.   Eucalyptus plantation (a) in the beginning and (b) after 3 years of receiving untreated sewage (Chhabra, 1989).

Fig. 7.7.   Two-year-old poplar plantation raised with untreated sewage water (Chhabra, 1989).

Table 7.13. Effect of disposing untreated sewage on the height, m, of different forest species

Treatments	Time, months							
	6	12	18	24	30	36	42	48
	Eucalyptus							
T_1	3.70	7.32	8.74	10.03	11.00	12.45	13.60	15.95
T_2	3.75	7.56	9.05	10.09	11.50	13.00	14.05	16.50
T_3	3.39	7.06	8.60	9.86	10.90	12.80	13.50	16.50
T_4	3.23	6.90	8.33	9.80	10.80	12.00	13.50	15.00
	Leucaena							
T_1	3.66	5.14	6.73	8.23	9.86	11.33		
T_2	4.24	6.08	7.84	9.44	10.85	12.08		
T_3	4.04	5.65	7.49	8.69	10.05	12.00		
T_4	4.03	5.40	7.22	8.22	9.85	10.70		
	Poplar							
T_1	6.19	8.90	11.95	13.90	15.11			
T_2	6.01	8.90	12.15	12.15	14.46			
T_3	5.71	7.98	11.00	11.00	15.56			
T_4	5.29	7.66	10.66	10.66	14.76			

T_1 = 15 cm SW daily, T_2 = 15 cm SW fortnightly,
T_3 = 15 cm SW monthly, T_4 = 15 cm tubewell water monthly.
Source : Chhabra (1989).

forest cover. Where such lands are not available, the same piece of land that is now being used as dumping ground for wastewaters can be shaped into ridges and furrows and used for plantation (Fig. 7.9).

7.7.4. Economic Viability

This technology is economically viable, is scale neutral, and does not require highly skilled personnel for its execution. In addition to saving the cost of treatment plants (initial as well as recurring expenditure), the system generates income from the sale of fuel wood or timber. The sludge accumulating in the furrows, along with the decaying forest litter, can be exploited as an additional source of revenue.

Reuse of wastewaters through irrigated forestry acts as a unique low-cost, pollution-free, and compatible wastewater treatment process and helps in mitigating the pollution of water bodies.It helps renovate and conserve the irrigation potential of these wastewaters right in the area of their production and is practical for developing countries in the arid and semi-arid regions.

7.8. Systems for Utilisation of Industrial Effluents

Effluents from the alcohol industry (Table 7.14), which uses molasses and malt as the raw material, have very high concentration of soluble salts, unfavourable

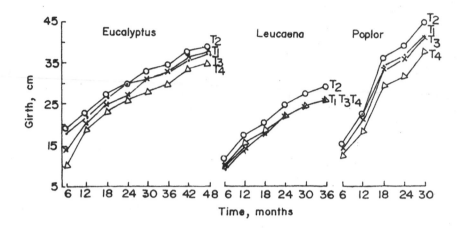

Fig. 7.8.   Effect of untreated sewage water on the girth of forest plant species (Chhabra, 1989).

Table 7.14.   Characteristics of the effluent from alcohol industry

Characteristic	Value
pH	7.8 to 8.5
EC, dS/m	10 to 14
Colour	dark brown
Odour	of burnt sugar
Temperature, °C	70–80
Total solids, mg/l	10,000–15,000
Suspended solids, mg/l	4,000–5,000
Inorganic matter, mg/l	40–60
Organic matter, mg/l	400-600
Free ammonia, mg/l	nil
BOD_5, at 20°C, mg/l	30,000–50,000
COD, at 20°C, mg/l	60,000–90,000
Na^+, me/l	4–6
K^+, me/l	70–80
Ca^{2+}, me/l	10–12
Mg^{2+}, me/l	50–60
CO_3^{2-}, me/l	nil-traces
HCO_3^-, me/l	50–60
Cl^-, me/l	40–50
SO_4^{2-}, me/l	20–30
P, mg/l	20–40
N (total), mg/l	60–80

Source: Chhabra (1991 b)

pH, high temperature, and high BOD$_5$ and COD (Chhabra, 1991 b). These effluents have a very unpleasant smell and a deep yellow to dark brown colour, which is normally due to lignin.

Effluents from the paper industry (Table 7.15) have medium to high concentration of soluble salts, high BOD$_5$ and COD, suspended solids, and toxic concentration of silica. These effluents are dark brown, mainly because of lignin (Chhabra, 1991c).

Table 7.15. Physicochemical characteristics of the effluent of a typical paper industry

Characteristic	Black liquor	Mixed treated effluent
Colour	Dark brown	Yellow
Smell	Mild	Mild
Clarity	Highly turbid	Transluscent
Temperature °C	33.5	25.5
BOD$_5$, mg/l	3500	40–100
COD, mg/l	5500	380–500
Total solids, mg/l	8000–10000	1000–1500
Organic matter, mg/l	4000–6000	200–500
Inorganic matter, mg/l	4000–5000	200–500
Free ammonia, mg/l	nil	nil
pH	8.0–9.0	6.50–7.00
EC, dS/m	5.0–5.5	1.15–1.50
Na$^+$, me/l	50–60	5–6
K$^+$, me/l	10–20	1–5
Ca^{2+}, me/l	0.75–1.0	1–3
Mg^{2+}, me/l	0.5–1.0	0.5–1.0
CO$_3{}^{2-}$, me/l	nil	nil
HCO$_3{}^-$, me/l	50–60	7–10
SO$_4{}^{2-}$, me/l	2–3	1–2
Cl$^-$, me/l	3–5	5–10
SAR, me/l	50–60	4–8
RSC, me/l	50–55	3–5
Toxic elements		
Pb, mg/l	nil	nil
Cd, mg/l	nil	nil
Ni, mg/l	nil	nil

Source: Chhabra (1991c).

Treatment of such waters in conventional treatment plant can decrease BOD$_5$, COD, and suspended solids and modify the temperature, but it cannot reduce the salinity level or degrade the persistent lignin. So such waters, even after treatment, are potential sources of soluble salts and can cause salinisation of soil and contamination of the groundwater. Hence, after pretreatment, these waters should be diluted with good quality waters so that the EC of the final water is within the permissible limit for a given soil-plant system. Since their volumes are quite large, a system based on forestry as described in section 7.7 can also be adapted for utilising such effluents.

8

Irrigation and Salinity Control

8.1. Basic Principles

The essential concern in farming practices for arid and semi-arid regions is to maintain, through some controlled irrigation schemes and within the constraints imposed by soil properties and water quality, the soil moisture stress at the lowest possible level. The overall soil moisture stress is the end product of: (a) the osmotic component, which is proportional to salt concentration, and (b) the soil matric component which is related to soil texture and the amount of irrigation water applied. In order to control these stress factors one requires: (a) appropriate leaching to achieve downward salt transport out of the rootzone so as to regulate osmotic pressure to the lowest practical level and (b) timely and adequate irrigation so as to maintain soil moisture stress within a range leading to the maximum crop yields. Therefore, the overall success of an irrigation programme should essentially be governed by an optimal combination of the following (interrelated) parameters:

   i)   Quality of irrigation water (to be considered in the light of soil properties and agronomic practices)

   ii)   Amount of irrigation water available.

   iii)   Drainage conditions

Before going into the quantitative aspect, the following basic aspects need to be considered:

   a)   Soluble salts imported into the soil by the irrigation water are concentrated in the soil by evaporation and transpiration. The resulting desiccation in the surface zone may generate a suction gradient in the surface layers that may cause upward movement of (salt bearing) soil water by capillary flow. Such a process is one of the main causes for salinisation in many soils, particularly in areas having high water table.

   b)   The soluble salt content in the rootzone will increase if the net downward movement of salt is less than the salt input from irrigation. Therefore, it is of key importance to keep the "salt balance" under control, and this is a function of the irrigation water (quality, quantity) and effectiveness of drainage.

c) Exchange reactions may occur between soil and irrigation water, and they may lead to gradual but detrimental changes in the drainage properties of the soil. Because of concentration effects with depth, ESP may attain levels that become critical in the lower part of the rootzone and have to be counteracted by addition of chemical amendments such as gypsum.

The Key-Processes are **IRRIGATION, LEACHING AND DRAINAGE.**

8.2. Salt Balance and Leaching Fraction

The leaching of soluble salts from the rootzone is a must in irrigated soils as without it, salt accumulation is bound to occur. The overall salt balance (SB) can be defined in terms of the various processes that contribute to the salt inflow and outflow, and the local changes in salt concentration in the soil. The SB can be written as a mass conservation equation:

$$SB = (DC)_{rain} + (DC)_{irr.} + (DC)_{ground} + S_{dissolv.}$$

$$- (DC)_{drain} - S_{prec.} - S_{crop} \tag{1}$$

where DC is the product of water depth (D) and salt concentration (C) for rain, irrigation water, groundwater, and drainage, respectively. S relates to the salt term originating from dissolution, precipitation, and uptake (crop) processes. Let us take the most simple case of a negligible contribution from dissolution and precipitation processes, crop removal, and groundwater effects. Equation (1) then takes the simple form of:

$$SB = C_{IW} \cdot D_{IW} - C_{DW} \cdot D_{DW} \tag{2}$$

Since it is convenient to express salt concentrations in terms of EC values, at steady-state conditions, i.e., when $SB = 0$, one should have

$$EC_{IW} \cdot D_{IW} - EC_{DW} \cdot D_{DW} = 0$$

$$\text{or } EC_{IW} \cdot D_{IW} = EC_{DW} \cdot D_{DW} \tag{3}$$

which states that, under the assumptions given, the amount of salt added during irrigation must be equal to the amount drained in order to maintain the salt balance. Perhaps the accuracy of this equation is less important than its conceptual value in that it states that leaching has to be ensured in irrigated agriculture, otherwise the salt concentration will rise to detrimental levels. The advantage of this equation is that it affords simple relationships for the water requirement of crops and the extent of leaching required, as related to the salt content (EC) of the irrigation water. Salinisation will occur if SB > 0 or if

$$EC_{IW} \cdot D_{IW} > EC_{DW} \cdot D_{DW} \qquad (4)$$

It is customary to take for the EC values some average for the period considered (a year or a season) and to take the annual amounts of IW and DW. Also if rains of short duration occur, then the average value for the IW can be corrected accordingly. As an example let us, estimate the rate of salinisation in the absence of rain and leaching. This rate can be expressed in terms of the directly measurable effects in laboratory tests such as SP, since this is the accepted way of quantifying soil salinity. The annual salinity increase can be written as:

$$\Delta ECe = \frac{(EC_{IW})(D_{IW}) - (EC_{DW})(D_{DW})}{D_{soil} \times \rho_{soil} \times SP/100} \qquad (5)$$

where D_{soil} is the soil depth concerned, ρ_{soil} is the bulk density (i.e., $D_{soil} \times \rho_{soil}$ = weight of soil); consequently, the denominator expresses the amount of water (receiving the EC increment) at SP conditions. Taking $D_{DW} = 0$, i.e., when there is no leaching, $EC_{IW} = 1,000$ μS/m, $\rho = 1.3$ g/cm^3 and SP = 40, we have

$$\Delta EC = \frac{1000 \times D_{IW}}{1.3 \times 0.4 \times D_{soil}} \qquad (6)$$

It is observed that for $D_{IW}/D_{soil} = 1$ (or a water gift of 30 cm for the top 30 cm soil) the increase in EC is already equal to about 2,000 μS/m. For $D_{IW}/D_{soil} = 2$, or an irrigation gift of 60 cm, the soil would turn saline (4 dS/m) in one season and that too with irrigation water of reasonably good quality. Consequently, the rate of salinisation is directly dependent on the amount and the salt content of the IW when some constant (average) annual amount of IW of specified EC value is being applied (Fig. 8.1). To remove these accumulated salts or to prevent their accumulation, the soil needs to be leached with excess water. The amount of this excess water required to leach the salts is known as leaching fraction (LF) and is defined as the ratio between the amount of water drained below the rootzone and the amount of irrigation water applied or the fraction of applied water that passes through the rootzone.

$$LF = \frac{D_{DW}}{D_{IW}} \qquad (7)$$

Of course, under-steady state conditions (i.e., when input = output, and no other removal or resolubilisation processes occur), and on the basis of equation (1), the value of EC_{DW} can be related to LF.

$$LF = \frac{D_{DW}}{D_{IW}} = \frac{EC_{IW}}{EC_{DW}} \qquad (8)$$

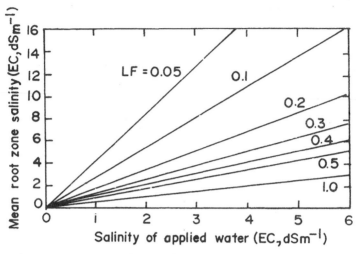

Fig. 8.1.   Mean rootzone salinity of soil as affected by EC of applied water at different leaching
fractions (adapted from Hoffman and Van Genuchten, 1983).

LF can be varied at will and, therefore, identical values of EC $_{DW}$ may be
obtained even when using waters of increasing EC values on the condition
that the LF is increased or, more precisely, that the ratio EC $_{IW}$/LF is kept
constant. Thus, to keep the rootzone salinity (EC $_{DW}$) at a constant level, the
EC $_{IW}$/LF ratio must be kept constant. Hence, doubling EC $_{IW}$ requires a
doubling of LF to obtain identical EC $_{DW}$. This is illustrated by the data in Fig.
8.2 and is based on the results of the lysimeter experiments. It is apparent
that, under steady-state conditions, salinity gradually increases from a level at
the soil top (controlled by EC $_{IW}$) to some higher level at the bottom; this salinity
level can therefore be controlled by manipulating leaching fraction.

8.3. The Leaching Requirement

The leaching requirement (LR) is the calculated fraction (depth) or quantity of
water that must pass through the rootzone to maintain the EC of the drainage
water at or below some specified level (EC'$_{DW}$). Such calculations are based
on equation (8):

$$LR = \frac{EC_{IW}}{EC'_{DW}} = \frac{D'_{DW}}{D_{IW}} \tag{9}$$

Equation (8) and equation (9) are the same except that, in equation (9),
EC'$_{DW}$ denotes the maximum permissible EC of the drainage water to maintain
crop yields, EC'$_{DW}$ being dependent on the nature of the crop, season, and
management factors. Leaching fraction is, therefore, the fraction of irrigation

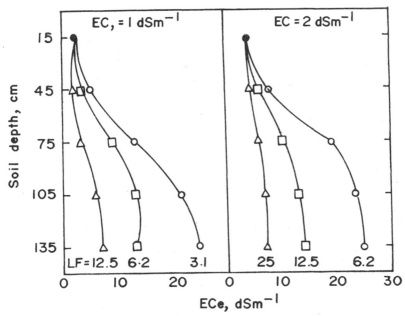

Fig. 8.2. Salinity distribution in the soil profile as related to the EC of IW and the leaching fraction (Hoffman, 1980).

water that passes through the lower rootzone, whereas the LR is a calculated estimate of the amount of leaching required to keep control of the soil salinity within the permissible limits.

8.3.1. LR as a Function of Quality of Irrigation Water

It is observed that LR depends on EC_{IW} (the higher EC_{IW}, the higher the LR) and EC'_{DW}, is the crop dependent threshold value which has to be specified (higher values being taken for salt-tolerant crops). Possibly, EC'_{DW} may be changed over the season if crop sensitivity varies with age.

The above concept can be illustrated by taking an example. Consider a crop that can tolerate EC'_{DW} of 8 dS/m. So if the soil is being irrigated with waters of EC 1, 2, and 3 dS/m, respectively, then the application of equation (9) leads to:

LR = 1/8 or 13% for EC_{IW} of 1 dS/m
2/8 or 25% for EC_{IW} of 2 dS/m
3/8 or 38% for EC_{IW} of 3 dS/m

8.3.2. LR as a Function of Water Consumptive Use

When using the LR concept for estimating the depth of irrigation water that must be applied (or the amount to be drained through the rootzone), information is needed on the amount of water required for consumptive use (D_{CW}), i.e., evapotranspiration or crop demand. In the absence of contribution from rain, we have

$$D_{IW} = D_{CW} + D_{DW} = D_{CW} + D_{IW} \cdot LR \tag{10}$$

$$D_{IW} = \frac{D_{CW}}{1-LR} \tag{11}$$

or

$$D_{IW} = D_{CW} \left(\frac{EC'_{DW}}{EC'_{DW} - EC_{IW}} \right) \tag{12}$$

However, it must be emphasized again that this equation is based on the assumption of uniform application of steady flow rate, no rainfall, and no salt removal by crop or precipitation.

8.3.3. LR and Drainage Capacity

The excess irrigation water to be applied (above the quantities needed for consumptive use) for leaching salts must of course not exceed the drainage capacity of the soil (cm/year). It is of interest to derive an expression relating LR and D_{DW}. Starting from equation (10), we may write:

$$D_{DW} = D_{IW} \cdot LR$$

In this, substituting the value of

$$D_{IW} = \frac{D_{CW}}{1 - LR} \qquad \text{(from equation 11)}$$

we get

$$D_{DW} = \frac{D_{CW}}{1 - LR} \cdot LR$$

$$D_{DW} = D_{CW} \cdot \frac{LR}{1-LR} \tag{13}$$

This equation can also be expressed in terms of EC_{IW} and EC'_{DW}

$$D_{DW} = D_{CW} \left(\frac{EC_{IW}}{EC'_{DW} - EC_{IW}} \right) \tag{14}$$

For irrigation water of different EC and a crop that can tolerate EC'_{DW} 8 dS/m, the D_{DW} will be:

$$EC_{IW}\ 0.5 \rightarrow D_{DW} = D_{CW}\left(\frac{0.5}{8 - 0.5}\right) = 6.66\% \text{ of CW}$$

$$1.0 \rightarrow D_{DW} = D_{CW}\left(\frac{1}{8 - 1}\right) = 14.28\% \text{ of CW}$$

$$2.0 \rightarrow D_{DW} = D_{CW}\left(\frac{2}{8 - 2}\right) = 33.33\% \text{ of CW}$$

$$4.0 \rightarrow D_{DW} = D_{CW}\left(\frac{4}{8 - 4}\right) = 100\% \text{ of CW}$$

Thus, when very poor waters are used and one wants to maintain EC of the drain water at a required threshold concentration, then the drainage capacity itself becomes a limiting factor. Also it, may not be possible to pass such a high amount of water when permeability is a limiting factor.

8.4. Factors Affecting Leaching Requirement

8.4.1. Rainfall

In many areas, rainfall may contribute a substantial amount of water and thus reduce the need of irrigation water required for leaching the salts. Under such situations, its contribution should be accounted for. If rainfall is sparse and spread over extended periods, a weighted average may be used for EC_{IW}:

$$\overline{EC}_{IW} = \frac{EC_{IW}\ D_{IW} + EC_{RW}\ D_{RW}}{D_{IW} + D_{RW}} \tag{15}$$

If, however, rainfall is confined to a short period, as in the monsoon type of climate, and that too during non-cropping season, then the rainfall itself may be sufficient to leach the salts out of the rootzone (Fig. 8.3). In such cases, very little leaching is required to control salinity due to the use of marginal quality waters in the early stages of the irrigation season.

8.4.2. Salt Precipitation

Another process that may lead to lower LR is salt precipitation. Lysimeter studies on the effect of LF on salt outflows in sub-surface drainage waters have shown that, in the range of 0.1 to 0.3, a LF of 0.1 would normally meet the LR for maximum crop production. However, at this low LF, it is generally not possible to compute an apparent salt balance from the values of salt inflow, salt outflow, and the salt content of the rootzone. The decreased outflow of salts at low LF is attributed to precipitation of calcium carbonate and possibly gypsum. This concentration-related precipitation of salts (in the soil) is greater

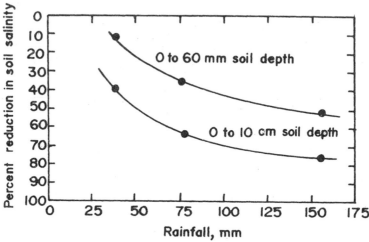

Fig. 8.3.   Effect of rainfall on the salinity of soil (Fang *et al.*, 1978)

if LF value of 0.1 or even lower can be achieved as compared to more usual values of 0.2 to 0.3.

Under such a situation, low LF will produce low volumes (quantities) of sub-surface drainage effluent and low salt outflow (quality) and ultimately low salt loading of river or drain water. Evidently, both these parameters, i.e., quantity and quality of sub-surface drainage water, are important factors affecting subsequent reuse of these effluents for irrigation purposes.

At high LF values, the "salt burden" to the sub-surface drainage effluent is ascribed to mineral dissolution, especially in calcareous and gypsiferous soils. The dissolution of minerals in the soil is more pronounced when high quality (low EC) irrigation waters are used for leaching and or those having low CO_3^{2-} and SO_4^{2-} as compared to Cl^- concentration.

8.4.3. Salt Uptake by Plants

Salt uptake by plants is generally insufficient to maintain the salt balance and reduce the leaching requirement. However, forage crops are able to take up a significant fraction of the salt input, particularly when the irrigation water is "non-saline". An average forage crop yielding 20 t/ha and having a salt (mineral) content of 5 per cent can remove up to 1 t of salts per year.

Certain halophytes and bushes can remove still higher amounts of salts. However, this biological harvesting of salts depends mainly on the high degree of salt tolerance of the crop itself. Certain plant species can be used for selective uptake of toxic elements such as Na, B, and Se and thus decrease the specific leaching requirement.

8.4.4. Salinity of Surface Layers

Many studies have shown that strict adherence to the assumption of steady-state salt balance may not be necessary and that salt accumulation can take place for short periods of time in the lower rootzone without affecting the crop yield. This can take place as long as salt balance is achieved over a long period of time and the crop is adequately supplied with water in the upper rootzone, where the major water use occurs (Fig. 8.4). Plants compensate for reduced water uptake from the highly saline zone by increasing its uptake from the low salinity zone . Reducing the LF has only a small effect on the salinity of this upper rootzone, since this area is adequately leached during each irrigation. However, under such situations, the salinity of the lower rootzone becomes greater, thus changing the salt concentration in the drainage water.

8.4.5. Method and Frequency of Irrigation

Uniform application of irrigation water with high frequency sprinkler or drip irrigation and even with well-managed surface irrigation system normally increases the efficiency of leaching and thus lowers the LR, provided that the interval between irrigations is not too long. The irrigation interval is an important factor since the crop must respond to the force with which water is held in the soil and also to the osmotic effects caused by salinity, both of which vary over time (Fig. 2.18). As the irrigation interval becomes greater, the osmotic effect becomes more dominant, especially when major water use begins to occur in

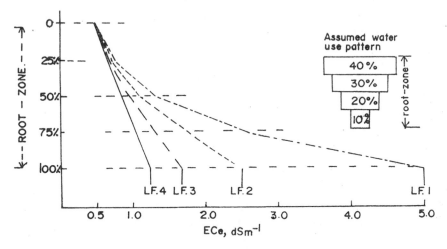

Fig. 8.4.   Salinity profile (ECe) expected to develop after long term use of water of 1 dS/m at four LF values and its effect on the pattern of water use by the plants (Shalhevet and Bernstein,1968).

the lower rootzone. This would become even more critical when poor quality water is used.

8.5. Calculations of the Leaching Requirement

From the quality of irrigation water, mode of irrigation, and the threshold salinity for a given crop, the LR can be calculated as per the equations discussed in the preceding sections. The LR value is a theoretical amount of leaching water needed to control salts in the rootzone and is based on field and laboratory experiences. However, the actual amount of water required will depend upon the efficiency of leaching, agronomic practices, and proper field management.

To calculate LR, for surface irrigation (including sprinkler), following steps are required :

a) Obtain EC_{IW} from the water analysis.

b) Obtain ECe value from Table 8.1 for a given crop to the degree of yield reduction (usually 10% or less). It is recommended that the ECe value for a 10 per cent yield reduction be used for field application since factors other than salinity are limiting yields greater than this in most instances. Values for yield reduction less than 10 per cent can be used if experience shows that near optimum yields can be obtained under the existing management conditions.

c) Calculate the LR by the expression

$$LR = \frac{EC_{IW}}{5\,EC_e - EC_{IW}} \qquad (16)$$

where LR refers to the minimum leaching requirement to control salts with ordinary surface irrigation methods.

For high frequency sprinkler or drip irrigation, LR is calculated through the following steps:

a) Obtain EC_{IW} from the water analysis.

b) Obtain the maximum ECe value from Table 8.1 for the given crop (100% yield loss).

c) Calculate the LR by the expression:

$$LR = \frac{EC_{IW}}{2\,(\text{max} \cdot EC)} \qquad (17)$$

The factor of 2 is obtained from EC_{SW}, which is equal to 2ECe.
Once the crop evapotranspiration demand (ET) and the desired LR are known, the net water requirement is derived from equation (18).

$$\text{Net water requirement} = \frac{ET}{1 - LR} \qquad (18)$$

where LR is expressed as a fraction.

Table 8.1.   Crop tolerance values (yield decrement to be expected for certain crops due to salinity of irrigation water when common surface irrigation methods are used)

Crop	0%		10%		25%		50%		Max.
	ECe^1	EC^2_{IW}	ECe	EC_{IW}	ECe	EC_{IW}	ECe	EC_{IW}	ECe^3
FIELD CROPS									
Barley[4]	8.0	5.3	10.0	6.7	13.0	8.7	18.0	12.0	28.0
(*Hordeum vulgare*)									
Cotton	7.7	5.1	9.6	6.4	13.0	8.4	17.0	12.0	27.0
(*Gossypium hirsutum*)									
Sugarbeet[5]	7.0	4.7	8.7	5.8	11.0	7.5	15.0	10.0	24.0
(*Beta vulgaris*)									
Wheat[6]	6.0	4.0	7.4	4.9	9.5	6.4	13.0	8.7	20.0
(*Triticum aestivum*)									
Safflower	5.3	3.5	6.2	4.1	7.6	5.0	9.9	6.6	14.5
(*Carthamus tinctorium*)									
Soybean	5.0	3.3	5.5	3.7	6.2	4.2	7.5	5.0	10.0
(*Glycine max*)									
Sorghum	4.0	2.7	5.1	3.4	7.2	4.8	11.0	7.2	18.0
(*Sorghum bicolor*)									
Groundnut	3.2	2.1	3.5	2.4	4.1	2.7	4.9	3.3	6.5
(*Arachis hypogaea*)									
Rice (paddy)	3.0	2.0	3.8	2.6	5.1	3.4	7.2	4.8	11.5
(*Oryza sativa*)									
Sesbania	2.3	1.5	3.7	2.5	5.9	3.9	9.4	6.3	16.5
(*Sesbania microcarpa*)									
Corn	1.7	1.1	2.5	1.7	3.8	2.5	5.9	3.9	10.0
(*Zea mays*)									
Flax	1.7	1.1	2.5	1.7	3.8	2.5	5.9	3.9	10.0
(*Linum usitatissimum*)									
Broad bean	1.6	1.1	2.6	1.8	4.2	2.0	6.8	4.5	12.0
(*Vicia faba*)									
Cowpea	1.3	0.9	2.0	1.3	3.1	2.1	4.9	3.2	8.5
(*Vigna sinensis*)									
Bean	1.0	0.7	1.5	1.0	2.3	1.5	3.6	2.4	6.5
(*Phaseolus vulgaris*)									
FORAGE CROPS									
Tall wheat grass	7.5	5.0	9.9	6.6	13.3	9.0	19.4	13.0	31.5
(*Agropyron elongatum*)									
Wheat grass (fairway)	7.5	5.0	9.0	6.0	11.0	7.4	15.0	9.8	22.0
(*Agropyron elongatum*)									
Bermuda grass[7]	6.9	4.6	8.5	5.7	10.8	7.2	14.7	9.8	22.5
(*Cynodon dactylon*)									
Barley (hay)[4]	6.0	4.0	7.4	4.9	9.5	6.3	13.0	8.7	20.0
(*Hordeum vulgare*)									

(Contd.)

Table 8.1.   *(Contd.)*

Crop	0%		10%		25%		50%		Max.
	ECe[1]	EC[2]$_{IW}$	ECe	EC$_{IW}$	ECe	EC$_{IW}$	ECe	EC$_{IW}$	ECe[3]
Perennial rye grass (*Lolium perenne*)	5.6	3.7	6.9	4.6	8.9	5.9	12.2	8.1	19.0
Trefoil birdsfoot, narrow leaf (*L. corniculatus tenuifolius*)	5.0	3.3	6.0	4.0	7.5	5.0	10.0	6.7	15.0
Harding grass (*Phalaris tuberosa*)	4.6	3.1	5.9	3.9	7.9	5.3	11.1	7.4	18.0
Tall fescue (*Festuca elatior*)	3.9	2.6	5.8	3.9	8.6	5.7	13.3	8.9	23.0
Crested wheat grass (*Agropyron desertorum*)	3.5	2.3	6.0	4.0	9.8	6.5	16.0	11.0	28.5
Vetch (*Vicia sativa*)	3.0	2.0	3.9	2.6	5.3	3.5	7.6	5.0	12.0
Sudan grass (*Sorghum sudanense*)	2.8	1.9	5.1	3.4	8.6	5.7	14.4	9.6	26.0
Wild rye beardless (*Elymus triticoides*)	2.7	1.8	4.4	2.9	6.9	4.6	11.0	7.4	19.5
Trefoil big (*Lotus uliginosis*)	2.3	1.5	2.8	1.9	3.6	2.4	4.9	3.3	7.5
Alfalfa (*Medicago sativa*)	2.0	1.3	3.4	2.2	5.4	3.6	8.8	5.9	15.5
Lovegrass (*Eragrostis* spp.)	2.0	1.3	3.2	2.1	5.0	3.3	8.0	5.3	14.0
Corn (forage) (*Zea mays*)	1.8	1.2	3.2	2.1	5.2	3.5	8.6	5.7	15.5
Clover, berseem (*Trifolium alexandrinum*)	1.5	1.0	3.2	2.1	5.9	3.9	10.3	6.8	19.0
Orchard grass (*Dactylis glomerata*)	1.5	1.0	3.1	2.1	5.5	3.7	9.6	6.4	17.5
Meadow foxtail (*Alopecurus pratensis*)	1.5	1.0	2.5	1.7	4.1	2.7	6.7	4.5	12.0
Clover, alsike, ladino, red, rawberry (*Trifolium* spp.)	1.5	1.0	2.3	1.6	3.6	2.4	5.7	3.8	10.0

FRUIT CROPS

Crop	0%		10%		25%		50%		Max.
Date palm (*Phoenix dactyliferia*)	4.0	2.7	6.8	4.5	10.9	7.3	17.9	12.0	32.0

(Contd.)

Table 8.1. *(Contd.)*

Crop	0%		10%		25%		50%		Max.
	ECe[1]	EC[2]$_{IW}$	ECe	ECIW	ECe	EC$_{IW}$	ECe	EC$_{IW}$	ECe[3]
Fig (*Ficus carica*),	2.7	1.8	3.8	2.6	5.5	3.7	7.4	5.6	14.0
Olive (*Olea europaea*)	2.7	1.8	3.8	2.6	5.5	3.7	7.4	5.6	14.0
Grape fruit (*Citrus paradisi*)	1.8	1.2	2.4	1.6	3.4	2.2	4.9	3.3	8.0
Orange (*Citrus sinensis*)	1.8	1.2	2.4	1.6	3.4	2.2	4.9	3.3	8.0
Lemon (*Citrus limonea*)	1.7	1.1	2.3	1.6	3.2	2.2	4.8	3.2	8.0
Apple (*Pyrus malus*)	1.7	1.1	2.3	1.6	3.3	2.2	4.8	3.2	8.0
Pear (*Pyrus communis*)	1.7	1.1	2.3	1.6	3.3	2.2	4.8	3.2	8.0
Walnut (*Juglans regia*)	1.7	1.1	2.3	1.6	3.3	2.2	4.8	3.2	8.0
Peach (*Prunus persica*)	1.7	1.1	2.2	1.4	2.9	1.9	4.1	2.7	6.5
Apricot (*Pyrus armeniaca*)	1.6	1.1	2.0	1.3	2.6	1.8	3.7	2.5	6.0
Grapes (*Vitis* spp.)	1.5	1.0	2.5	1.7	4.1	2.7	6.7	4.5	12.0
Almond (*Prunus amygdalus*)	1.5	1.0	2.0	1.4	2.8	1.9	4.1	2.7	7.0
Plum (*Prunus domestica*)	1.5	1.0	2.1	1.4	2.9	1.9	4.3	2.8	7.0
Blackberry (*Rubus* spp.)	1.5	1.0	2.0	1.3	2.6	1.8	3.8	2.5	6.0
Boysenberry (*Rufus* spp.)	1.5	1.0	2.0	1.3	2.6	1.8	3.8	2.5	6.0
Avocado (*Persea americana*)	1.3	0.9	1.8	1.2	2.5	1.7	3.7	2.4	6.0
Raspberry (*Rubus idaeus*)	1.0	0.7	1.4	1.0	2.1	1.4	3.2	2.1	5.5
Strawberry (*Fragaria* spp.)	1.0	0.7	1.3	0.9	1.8	1.2	2.5	1.7	4.0

VEGETABLE CROPS

Crop	0%		10%		25%		50%		Max.
Beet[5] (*Beta vulgaris*)	4.0	2.7	5.1	3.4	6.8	4.5	9.6	6.4	15.0
Broccoli (*Brassica italica*)	2.8	1.9	3.9	2.6	5.5	3.7	8.2	5.5	13.5
Tomato (*Lycopersicon esculentum*)	2.5	1.7	3.5	2.3	5.0	3.4	7.6	5.0	12.5

(Contd.)

Table 8.1.   *(Contd.)*

Crop	0%		10%		25%		50%		Max.
	ECe[1]	EC[2]$_{IW}$	ECe	ECIW	ECe	EC$_{IW}$	ECe	EC$_{IW}$	ECe[3]
Cucumber (*Cucumis sativus*)	2.5	1.7	3.3	2.2	4.4	2.9	6.3	4.2	10.0
Cantaloupe (*Cucumis melo*)	2.2	1.5	3.6	2.4	5.7	3.8	9.1	6.1	16.0
Spinach (*Spinacia oleracea*)	2.0	1.3	3.3	2.2	5.3	3.5	8.6	5.7	15.0
Cabbage (*Brassica oleracea capitata*)	1.8	1.2	2.8	1.9	4.4	2.9	7.0	4.6	12.0
Sweet corn (*Zea mays*)	1.7	1.1	2.5	1.7	3.8	2.5	5.9	3.9	10.0
Sweet potato (*Ipomea batatas*)	1.5	1.0	2.4	1.6	3.8	2.5	6.0	4.0	10.5
Pepper (*Capsicum frutescens*)	1.5	1.0	2.2	1.5	3.3	2.2	5.1	3.4	8.5
Lettuce (*Lactuca sativa*)	1.3	0.9	2.1	1.4	3.2	2.1	5.2	3.4	9.0
Radish (*Raphanus sativa*)	1.2	0.8	2.0	1.3	3.1	2.1	5.0	3.4	9.0
Onion (*Allium cepa*)	1.2	0.8	1.8	1.2	2.8	1.8	4.3	2.9	7.5
Carrot (*Daucus carota*)	1.0	0.7	1.7	1.1	2.8	1.9	4.6	3.1	8.0

1   ECe means electrical conductivity (EC) of the saturation extract paste in dS/m.
2   EC$_{IW}$ means EC of the IW in dS/m at 25°C. This assumes about a 15–20% leaching fraction and an average salinity of soil water taken up by crop about three times that of the irrigation water applied ($EC_{SW} = 3\ EC_{IW}$) and about twice that of the soil saturation extract ($EC_{SW} = 2ECe$). From the above, $ECe = 3/2\ EC_{IW}$. New crop tolerance table for EC_{IW} can be prepared for conditions that differ greatly from those assumed in the guidelines. The following are estimated relationships between EC_e and EC_{IW} for various leaching fractions: LF = 10% ($ECe= 2EC_{IW}$); LF 30% ($ECe = 1.1EC_{IW}$), and LF = 40% ($ECe= 0.9\ EC_{IW}$).
3   Maximum ECe means the maximum EC of the soil saturation extract that can develop because of the listed crop drawing soil water to meet its evapotranspiration demand. At this salinity, crop growth ceases (100% yield decrement) because of the osmotic effect and reduction in crop water availability to zero.
4   Barley and wheat are less tolerant during germination and seedling stage. ECe should not exceed 4 or 5 dS/m.
5   Sensitive during germination. ECe should not exceed 3 dS/m for gardenbeet and sugarbeet.
6   Tolerance data may not apply to new semi-dwarf varieties of wheat.
7   An average for Bermuda grass varieties. Suwanee and Coastal are about 20% more tolerant; Common and Greenfield are about 20% less tolerant.

Source: Ayers and Westcot (1985).

Rainfall may meet a portion of the crop ET demand or it may accomplish part or all of the needed leaching. This will depend upon soil conditions and rainfall patterns and is to be taken into consideration while determining net water requirements.

The LR calculated above should be adequate to control salts unless they are already present in excess of crop tolerance, in which case an initial heavy leaching may be required to remove accumulated salts. After such initial leaching, full production potential may be restored.

Example

Sorghum is being irrigated by basins on a uniform loam soil using Pecos River water, $EC_{IW} = 3.2$ dS/m. With an ET demand of 5 mm/day and irrigation applied after every 20 days, 100 mm of water would be used. If the application efficiency is 0.65, then 100/0.65 = 155 mm of water must be applied with each irrigation to meet crop ET demand. How much additional water should be applied for leaching?

Step 1

Given: $EC_{IW} = 3.2$ dS/m

$ECe = 5.1$ dS/m (Table 8.1 for sorghum at 10% yield loss)

$$LR = \frac{3.2}{5\,(5.1) - 3.2} = 0.14$$

Step 2

Given: $ET = 100$ mm/irrigation

$LR = 0.14$ (from step 1)

$$\text{Net water requirement} = \frac{100}{1 - 0.14} = 116 \text{ mm/irrigation}$$

The deep percolation losses (55 mm) are larger than the leaching requirement. If the deep percolation losses are assumed to be uniform and no runoff occurs, then there is no need to add the leaching requirement to the unavoidable deep percolation losses. Uniform application of this water along with increasing efficiency of application should be encouraged.

The tables also include the suggested maximum ECe at which reduction in yield would be 100 per cent (if the entire rootzone is at this salinity, the crop cannot extract water and the growth stops). The maximum value of ECe is obtained by extrapolating the decrease from the 0-10-25-50 per cent yield reduction values to 100 per cent as illustrated in Fig. 8.5.

$$EC_{SW} = ECe \times 2$$

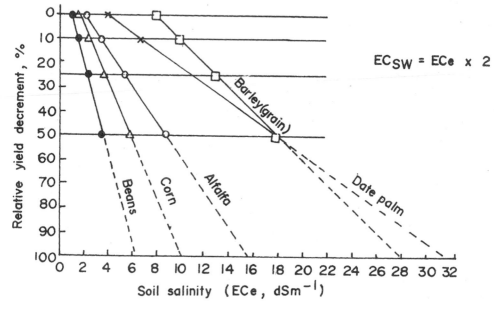

Fig. 8.5.   Method of determining maximum ECe for crop yield (Ayers and Westcot, 1985)

As shown in Fig. 8.5 each increase in soil salinity, i.e., ECe, in excess of the concentrations that initially begin to affect yield, will cause a proportional decrease in yield. This linear effect is expressed by the following equation:

$$Y = 100 - b(ECe - a) \qquad (19)$$

where Y is relative crop yield (%), a is the salinity threshold value for the crops representing the maximum ECe at which a 100 per cent yield can be obtained, and b is the yield decrement per unit salinity or per cent yield loss per unit of salinity (ECe) between the threshold value (a) and the ECe value for a 100 per cent yield decrement.

The crop tolerance tables were prepared using this formula on the basis of available values. The conversion from ECe to ECsw assumes a LF of 15–20 per cent. Other important assumptions in the tolerance tables are that yields are closely related to the average salinity of the rootzone and that the water uptake is normally much higher from the upper rootzone as assumed with the 40-30-20-10 per cent relationship in the guidelines.

It must be recognised that actual production with water of the quality indicated can range from the full 100 per cent potential down to zero, depending upon any one of many factors other than water quality. The values given in Table 8.1 represent the maximum production potential for the quality of water under optimum conditions of use.

The values suggested as tolerance limits to salinity of applied water (EC$_{IW}$) may seem high at first glance. However, when these suggested values are compared with those from field trials using relatively poor quality waters, there appears to be reasonably good agreement on salinity tolerance of crops tested.

Crop tolerance is presented in the tables as if tolerance were a fixed value. This is not exactly true. Crop tolerance does change with water management practices and is influenced by the stage of growth, nature of rootstock, variety, and climate. For many crops, such as sugarbeet, rice, wheat, barley and several other vegetables, the germination and early seedling stage is the most sensitive and soil salinity (EC$_e$) in excess of 4 dS/m around the germinating seed may delay or inhibit germination and early growth. The tolerance values as presented in Table 8.1 are based on the response from late seedling stage to maturity.

Rootstocks influence the salinity tolerance of certain tree crops such as citrus. Crop varieties, such as for grapes and almond, exhibit significant differences in salt tolerance. The differences in salinity tolerance have been used in making both rootstock and variety selection for commercial planting. Annual crops, too, show variation in response to salinity. Plant breeding and crop selection for salinity tolerance are now being emphasised and results are stimulating new research for genetic salt tolerance among varieties.

Climate plays an important role in crop tolerance. In general, crops will be more tolerant to adverse salinity as compared to during warmer periods and times of low humidity or high evapotranspiration.

8.6. Timing of Leaching Irrigation

The timing of leaching does not appear to be critical, provided crop salinity tolerance limit is not exceeded for extended periods of time or critical stage of plant growth. The leaching can be done at each irrigation, after a few irrigations, once a year, or after long intervals.

Providing leaching at each irrigation when the crop is growing is the best way to avoid salinity stress. It is highly beneficial for crops sensitive to salinity build-up even for a short period. However, in soil with low infiltration rate and for crops sensitive to excess moisture in the rootzone, leaching of soil at each irrigation may not be possible. Furthermore, during the cropping season, demand of water to meet the irrigation needs of crops in the command area may not allow for additional water for leaching of salts. Since the salinity tolerance increases with age and for those crops that can tolerate high salt concentration, leaching may be provided at the end of the cropping cycle. Leaching outside peak water use periods will also reduce the design capacity of the distribution system and may influence drainage design factors as well.

Regardless of the method used, soil and crop should be adequately monitored. Soil and plant tissue analysis can help determine need and timing of leaching. In most cases, an annual leaching during non-crop or dormant crop periods, as during the winter season, is preferred. Rainfall in some cases may be adequate to accomplish all the needed leaching.

Sometimes, soil conditions may prevent flexibility in mode of leaching. If soil infiltration rate is low, leaching needs to be postponed till cropping is over. The effects of fallow periods on soil salinisation also need to be considered. Water availability may also prevent flexibility, thus allowing only after harvest or presowing leaching or scheduling of leaching outside periods of peak water requirements.

8.7. Methods to Reduce Leaching Requirement

Maximising the efficiency of leaching or reducing the LR may reduce water needs. Leaching of the salts is maximum under unsaturated conditions. Hence any factor related to soil or method of application of irrigation water that affects unsaturated flow in the soil will decrease the wastage of water, reduce the leaching requirement, and provide maximum efficiency of leaching. In most cases, flexibility in the management choice may be limited but several management tips suggested here may apply to a certain irrigation situation:

a) Grow crops during the cool season instead of the warm season, since LR is related to the ET demand.

b) Grow salt-tolerant crops, to save on leaching demands.

c) Apply soil management practices that limit flow into and through large pores, such as tillage to reduce the number of surface cracks, root-holes, worm-holes, and other large pores.

d) Practice chiselling or subsoiling or deep tillage to increase permeability of lower layers.

e) Add amendments where permeability of the surface soil is a problem.

f) Use irrigation methods such as sprinklers that apply water below the infiltration rate of the soil, thus reducing water movement through large pores. This will require more irrigation time but uses less water than continuous ponding.

g) Wet the soil prior to the onset of winter rains where rainfall is insufficient for a complete leaching. Even a little rainfall on a wet soil is efficient in leaching since the rain moves deeper into the soil and provides high quality water to the upper rootzone.

h) Where drains are provided, leach in stages: first leach the area in the centre between drains, then leach closer to the drain.

Under field conditions, the above factors together with soil texture and depth of water table may affect the leaching efficiency. In the calculation of the actual amount of water required, leaching efficiency factor must be included for a given specific condition.

8.8. Salinity Control through Cultivation and Deep Tillage

Cultivation and deep tillage are also effective but temporary solutions for salinity control in soils with permeability problem. Cultivation loosens and roughens the surface soil but is usually done for reasons other than to improve water penetration. However, where penetration problems are severe, cultivation or tillage may be particularly helpful. A rough, cloddy furrow or field as compared to a smooth one will prevent runoff, and improve penetration for the first irrigation or two. A normal cultivation procedure can sometimes be modified to leave a rougher surface.

Deep tillage, also known as chiselling or subsoiling, physically tears, rips apart and shatters the soil at deeper depths. It is of particular help in those soils where hardpans formed due to accumulation of clay or precipitation of $CaCO_3$ or sodication due to high RSC or SAR subsoil waters act as physical barriers and limit the permeability of lower layers. Even though it is not a permanent solution, it may improve the situation enough to make an appreciable difference in crop yield. Deep tillage should be done prior to planting or during periods of dormancy when root pruning or root disturbance of permanent crops is less disruptive. Deep tillage is most effective when the soil is dry enough to shatter and crack. If done on wet soil, compaction, aeration, and permeability problems can increase and adversely affect the crop yield.

With low salinity waters ($EC_{IW} < 0.5$ dS/m) the permeability problem usually occurs in the upper few centimetres of soil. A surface crust or nearly impermeable surface soil is a typical phenomenon under such situations. Cultivation breaks this surface crust, roughens the soil and closes cracks and air spaces. This hastens the flow of water and greatly increases the surface area exposed for infiltration.

In contrast, the permeability problem due to high sodium waters (high adj. SAR) may occur initially near the surface but progressively extend to deeper layers as the season advances or from year to year. Under such situations cultivation and deep tillage may permit increased quantities of water to enter the soil, but usually only for a relatively short period.

8.9. Leaching and Reclamation

Traditionally, saline soils have been reclaimed by flooding or by ponding water. In general, the depth of soil leached is roughly equal to the depth of water infiltrated during leaching. Displacing 1 pore volume of soil water lowers the salt level by a factor of approximately 2. The displacement of 1.5 to 2 pore volumes reduces soil salinity by about 80 per cent. Since, the pore volume of soils is generally about 50 per cent the displacement of 2 pore volumes is about equal to an equivalent depth of water per unit soil depth. The reasoning, as given above, is based on early field data (obtained by ponding), the results of which are summarised in Fig. 8.6.

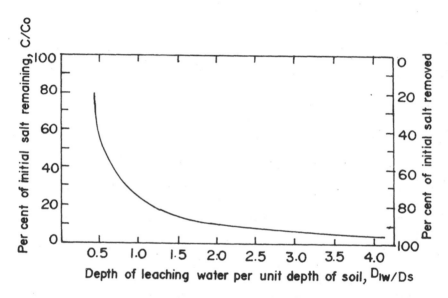

Fig. 8.6.   Leaching of soil as a function of depth of irrigation water (Hoffman, 1980).

These data form the basis of the old empirical rule that, for reclaiming saline soils, one foot of water must be supplied per foot of soil depth to remove 80 per cent of the soluble salts. This rule was derived on the basis of conditions where all the water was applied over a short time span, without any runoff.

More recent experiences (Fig. 8.7) have shown that reclamation based on the use of intermittent water application or sprinkling leads to a much more efficient salt removal. In some cases it is reported that intermittent ponding (which requires limited border strips, thereby reducing field preparation costs) can cut the water requirement for soil reclamation by a factor of two or more. Evidently, the higher yield in intermittent ponding or sprinkling is due to the fact that salt is displaced more efficiently under conditions of partial saturation. In such cases, the rate of water movement is much slower (being more restricted to the narrow pores) than that under saturated conditions (due to the higher velocity of water flow in macropores).

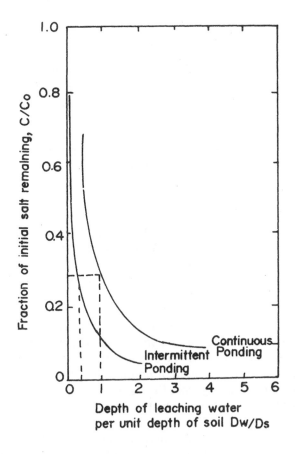

Fig. 8.7.   Effect of continuous and intermittent ponding on leaching efficiency of soil (Oster *et al.*, 1984).

8.10. Salinity Control and Methods of Irrigation

Under arid and semi-arid conditions, the major bottleneck in optimising agricultural production is the limited availability of good quality water. So efforts have to be made to:

a) find methods and techniques to increase the efficiency of limited quantities of good quality water, and

b) to exploit underground saline waters that in the past have been considered unsuitable for agriculture.

To achieve the above objectives, it is important to develop irrigation systems and improve application techniques in such a way that one can provide water at the right place, at the right time, in the right amounts for optimising plant growth without invoking salinity hazards to the soil. These aims can be achieved through localised irrigation like trickle or drip and porous tube irrigation, which is most efficient in terms of water distribution and application. It is easy to manage, and through it one can not only apply controlled amounts of water, but also uniformly provide nutrients, growth regulators, and insecticides. However, these systems, because of their high cost of installation and sophisticated mechanisation (mainly to save labour), are unsuitable for many developing countries. Furthermore, sprinkler and drip irrigation are not suited to all qualities of water and all soil conditions, climate, or corp. Several important factors should be considered before attempting to improve salinity control by changing the method of irrigation.

8.10.1. Surface Irrigation

Flood, basin, furrow, and border methods apply water at intervals to allow the crop to utilise 50 per cent or more of the available water in the rootzone before receiving the next irrigation. Since water is used by the crop during each irrigation interval the soil becomes drier (Fig. 2.18) and the soil water saltier at the end of each irrigation cycle. This adversely affects the availability of water to the crop and subsequently its yield.

The benefits of more frequent irrigations and routine leaching have been already emphasised. Surface irrigation methods are often not sufficiently flexible to allow adjustments in timing and depth of water. For example, it may not be possible to reduce the depth of water applied below 80 to 100 mm per irrigation. As a result, irrigating more frequently through surface irrigation may reduce salinity but may also waste water, cause waterlogging, and result in reduced yields.

In such cases, for better water and salinity management, a change in method of irrigation to sprinklers or drip may be needed. Such a change is costly and will require justification in terms of improved yields, better crop adaptability, or other benefits that can realistically be expected.

8.10.2. Sprinkler Irrigation

A good sprinkler system must meet the requirements of the crop for water (ET), of the soil as to rate of application and water storage capacity, and of the water and crop as to leaching requirement (LR). Special on-site conditions and peculiarities of crops, soils, water supply, or climate must also be considered.

With adequate system design and management, moveable and fixed sprinklers can apply water with good uniformity and with flow rates low enough to prevent runoff. This results in an excellent overall water supply to the crop and adequate and uniform leaching. Depth of water applied can also be controlled by adjustments in the duration of application.

Sprinklers are sometimes used to aid germination and early seedling growth, at which time the crops may be particularly sensitive to salinity, high temperatures, and soil crust. With solid set systems used for crop germination, irrigations are applied once or more each day for several days and for relatively short periods of 1 to 3 hrs. After 10 to 14 days the sprinklers are moved to another field and the process is repeated. In this way, a sprinkler system can be used for germinating in several fields in a season. With portable or wheel-roll system, irrigations are frequent enough to maintain low salinity and reduce soil problems such as hard crust.

Sprinklers often allow efficient and economic use of water and reduce deep percolation losses. If water application through sprinklers is in close agreement with crop needs (evapotranspiration and leaching), drainage and high water table problems can be greatly reduced, which in turn should improve salinity control.

When using poor quality waters, sprinklers do offer a hazard to sensitive crops. Fruit plants such as grapes, citrus, and most tree crops are sensitive even to relatively low concentrations of Na^+ and Cl^- present in the soil water and try to avoid their absorption through roots. However, these plants, under conditions of low humidity, may absorb (through foliage) excessive amounts of these toxic ions from the sprinkler-applied water that wets the leaves. The toxicity shows up as leaf burn (necrosis) on the outer edges and can be confirmed by leaf analysis. Irrigation during periods of high humidity, low temperature, and low transpiration rates, such as at night, has often greatly reduced or eliminated these problems. Annual crops are generally tolerant to moderate levels of Na^+ and Cl^- in the irrigation water. However, they may also be more sensitive to salts absorbed through the leaf during sprinkling than to similar water salinities when applied by surface or drip method.

The toxicity due to Cl^-, SO_4^{2-}, and HCO_3^- ions in sprinkler irrigation will be greater in close environments like greenhouses than in open fields. In the closed environment, after the evaporation of water, the salts continue to remain on the foliage and get absorbed; in the open fields, they may get washed away by the rains. In closed environments a concentration of Cl^- as low as 1.0 me/l of irrigation water may be toxic to sensitive ornamental plants like azaleas, camellia, rhododendrons and fruit nurseries. Where water salinity is severe, many trials should be conducted to test the suitability of sprinkling under existing local conditions. This may even be needed for crops not presently considered to be sensitive to specific ion toxicities.

8.10.3. Drip (Trickle) Irrigation

Drip irrigation is a method that supplies the required quantity of water to the crop almost on a daily basis. Water is applied from each of many small emitters at a low rate. The timing and duration of each irrigation can often be regulated by time clocks (or hand valves) with adjustments in water applied being made through the duration of irrigation, by changing the number of emitters, or both.

With good quality water, yield with drip irrigation should be equal to, or slightly better than, that obtained with other methods under comparable conditions. With poor quality water, yields may be better with drip irrigation because of the continuous high moisture contents and daily replenishment of water lost by evapotranspiration. Frequent sprinkler irrigation might also give similar results but the leaf burn and defoliation of sensitive species would not be expected with drip irrigation. If poor quality water is used and crop tolerances are exceeded by the usual methods of irrigation, a better yield may be possible with drip irrigation although yields may not be as high as those obtained while using good quality water. However, even with no expected yield benefits, other advantages such as possible savings in water, fertiliser or labour may be high enough in special cases to justify the added investment costs of the drip system.

With the drip method, salts do accumulate both at the soil surface and within the soil at the outside edges of the area wetted by the emitters. Salts may also accumulate below the emitters but the daily irrigations, if properly applied, should maintain a slight but nearly continuous downward movement of moisture to keep these salts under control. With time the salt accumulation at soil surfaces and in wetted fringe areas between emitters can become appreciable. Such accumulation is a hazard if it is moved by rain into the rootzone of the crop or, in the case of annual crops, if a new planting is made in these salty areas without prior leaching. If rainfall is sufficient each season to leach the accumulating salts, no problems are anticipated. However, if rainfall is insufficient or infrequent, problems may result. Leaching by sprinklers or surface flooding prior to planting has been effective in removing accumulated salts. This will require a second irrigation system and use of additional water but may allow continued production using poor quality water.

8.10.4. Pitcher Irrigation—a Sub-surface Irrigation Technique

A simple, efficient, and economic way to provide localised sub-surface irrigation known as "pitcher irrigation" was developed at the Central Soil Salinity Research Institute, Karnal, India (Mondal, 1974). In this technique, a baked earthen pot called as pitcher is used to provide and distribute water in the rootzone. The pitcher is buried in the soil and filled with water. Through the pores, water oozes out and wets the soil in the vicinity of the pitcher, thus making it fit for sowing seeds and planting seedlings (Fig. 8.8).

Water is filled in the pitcher manually, through a hose or by buckets, once or twice a week depending upon the water depletion rate, which in turn depends upon the type of crop, stage of growth, and the climatic conditions.

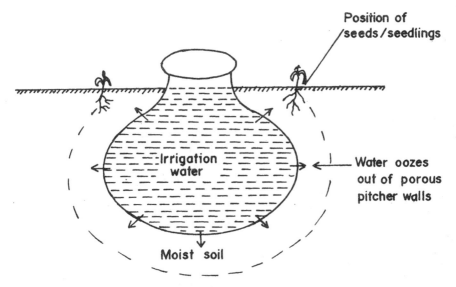

Fig. 8.8. Schematic diagram showing functioning of pitcher irrigation (Chhabra, 1990).

Fertilisers and insecticides are also added in the pitcher itself for uniform application. Various aspects of pitcher irrigation are discussed below.

a) Water discharge rates
The amount of water that can flow out through the walls of the pitcher, i.e., discharge per unit surface area per unit time, depends upon:
   i) The waterhead at a point: This is maximum at the bottom and thus move water flows through the bottom surface.
   ii) The apertures and intensity of pores in the earthen wall: Normally, the pores are quite small and thus the discharge is slow but enough to meet the needs of the plants. Within a reasonable limit, one can increase or decrease the size of apertures by varying the ratio of clay and sand, which are the basic materials used for making the pitchers.
   iii) Evapotranspiration need of the plant: It is greater during the day than at night. The water flow is affected by the suction created by the evapotranspiration of the plant.
   iv) Texture of the soil: The discharge is usually greater in a light-textered soil than in a heavy textured soil.

b) Wetting front
When the pitcher is filled, water starts coming out through the pores and wetting front advances both horizontally and vertically. However, the horizontal

movement is restricted and can be predicted by the following power curve, expressed as:

$$L = at^b \tag{20}$$

where L = length of advance from the centre of the pitcher,

   t = time in days, and

   a,b= are the empirical constants, the value of which depends upon soil type, evaporative demand, discharge of the distributor, and the salinity of the irrigation water. During the winter, value of a was 16.0 and for b for a 0.25. sandy loam soil.

Generally, for a pitcher of 30 cm diameter and 10 litres water holding capacity, wetted surface area is around 0.7 m^2. It is sufficient to grow at least four plants of any vegetable crop.

c) Moisture content of the soil
The moisture content of the soil around the pitcher varies from a maximum of 20 per cent near the pitcher wall to normal levels of dry soil, i.e., 5 to 10 per cent, at the end of the wetting front. The maximum moisture is nearly equal to field capacity of the soil and thus is most favourable for the growth of plants. Due to limited hydraulic conductivity of the pitcher wall, moisture levels above these are rarely obtained, except at night when, because of low evapotranspiration demand, the moisture percentage may slightly increase and encourage deep percolation losses. However, these losses can be minimised by compacting the soil or putting a plastic sheet just below the pitcher.

d) Salinity development in the soil
The pitcher irrigation method being essentially a sub-surface irrigation system, the salts accumulate at the soil surface and at the boundaries of the wetted front, leaving the rootzone in equilibrium with the salinity of the irrigation water used in the pitcher. Salt distribution at a distance of 15 cm from the wall of the pitcher as influenced by the quality of irrigation water presented in Table 8.2, shows that the salinity build-up is very low near the pitcher.

e) Plant response
Method and frequency of irrigation greatly modify the salinity effects. It has been generally observed that for the same amount of applied water, the salinity stress can be minimised by increasing the frequency of irrigation. In pitcher irrigation, since water is constantly being supplied to the rootzone, this minimises the matric stress and thus allows the plant to tolerate greater osmotic stress.

Table 8.2. Salt distribution, at 15 cm from the pitcher, as influenced by quality of irrigation water.

Depth, cm	ECe of soil for EC_{iw}, dS/m			
	4	8	12	15
0– 7.5	3.12	4.72	5.72	9.82
7.5–15.0	3.12	4.40	4.72	7.96
15.0–22.5	2.84	4.44	5.16	5.92
22.5–30.0	2.84	3.32	5.68	6.08

Source: Adapted from Dubey *et al.* (1988).

Using the Maas and Hoffman (1977) response function curve, it has been observed that plants grown with pitcher irrigation tolerate higher irrigation water salinity as compared to with surface system of irrigation (Table 8.3).

Table 8.3. Tolerance indices of vegetable crops using pitcher irrigation and surface irrigation methods

Crop	Pitcher irrigation		Surface irrigation	
	EC_t*	EC_{50}**	EC_t	EC_{50}
Cabbage	9.7	15.4	2.70	10.43
Brinjal	9.8	16.4	-	-

*EC_t refers to threshold conc. of salts up to which there is no decrease in yield.
**EC_{50} refers to conc. of salts at which there is 50 per cent reduction in yield.
Source: Dubey *et al.* (1988).

Though this technique can be used for raising most vegetables, fruit plants, and forest trees, it is ideal for species such as cucurbits which require more space for spreading their aerial parts but have roots confined to relatively smaller areas.

f) Life of the pitcher
Baked clay pitchers can be continuously used for 3 to 6 years. Clogging of pores due to precipitation of salts, as with the dripper, is a remote possibility when the pitcher is kept wet continuously. Clogging of pores from outside due to salts from the soil is limited because the flux is always continuous and occurs outwards, i.e., away from the wall of the pitcher. The plant roots can neither enter the pitcher by penetrating through the pores nor seal the surface from outside. They also do not exert enough pressure to break the wall of the pitcher.

To prolong the life of the pitcher, the following precautions should be taken:
i)   While not irrigating, keep the mouth of the pitcher covered. This will minimise the loss of water as well as prevent sunlight from entering the pitcher and minimise algal formation and growth.

ii) Only clean water should be used for filling the pitcher. Muddy runoff from rainfall should be used after it is passed it through a sand filter.

iii) Before storing the pitchers, wash them with good quality water so as to remove the salts and to prevent their precipitation and subsequent clogging of the pores on drying.

g) Applicability to other situations

Pitcher irrigation is a most effective and low-cost technology for using a limited supply of good quality water and also for using saline water resources. It is most suited to the following situations:

i)   Where availability of good quality water is scarce.

ii) Undulating areas where soils are difficult to level for uniform application of irrigation waters.

iii) Light soils with high infiltration rate, which cannot be efficiently irrigated by surface method.

iv) Areas with highly saline water sources, which cannot normally be used through surface method of irrigation.

Though this technique is costlier than surface irrigation system, it is worth the additional cost in view of the easy labour availability in many developing countries and the saving of water it provides.

8.11. Solutions for Problems of Disposing Drainage Water

The traditional drainage system lowers the water table, minimises upward flow and thus build-up of salts due to capillary rise, and also provides for adequate leaching. The drainage water collected is normally discharged into an open surface drain and then into the sea or a canal/river irrigation system. But sometimes the disposal of drainage water, rich in salts and toxic elements such as B and Se, can cause problems in downstream areas for soil, human and animal health.

To minimise the disposal problem, the volume of drainage water must be reduced. This can be achieved by:

a) Use of more efficient techniques for leaching, i.e., intermittent ponding rather than continuous ponding of water.

b) Improvement in efficiency of irrigation by choosing the appropriate method of irrigation and land configuration.

c) Use of less water-demanding crops, thereby avoiding frequent irrigation and eventually the need of leaching the accumulated salts.

d) Reuse of the drainage water for:

    i) initial leaching of adjacent highly saline soils,

    ii) irrigating salt-tolerant crops,

    iii) consumptive use with good quality canal water, and

    iv) brackish fish pond culture.

Reuse of drainage water helps not only to increase the overall water supply and thus increase crop production, but also to reduce the volume of drainage water. This in turn helps decrease the capacity of the drainage conveyance and treatment system and the salt load of the receiving river water.

Saline drainage water can be used, for crop production just like any other saline water (Fig. 8.9), by:

a) Irrigating with good quality water during critical stages of crop growth (like germination, seedling establishment, and flowering) and with saline water in between.

b) Using good quality water for sensitive crops and drained saline water of that area in the next part for semi-tolerant and then for tolerant crops. This way, even grasses or halophytes are grown with highly saline water.

c) Raising only one semi-tolerant crop with diluted saline water and allowing the second half of the year (normally monsoon period) for desalinisation through rains and/or using limited good quality water for critical stage of the crops.

Using these concepts, Sharma *et al.* (1989) reported that under drainage system, reuse of drainage water having salinity of 6, 9, and 12 dS/m, with limited quantity of good canal water (for presowing irrigation) can support a good crop of wheat (Table 8.4) over a long period in a mono-cropped area.

Table 8.4. Effect of different salinity levels of the drainage water on yield of wheat under drainage system

Treatment	Wheat grain yield, t/ha				% reduction in yield over canal water
	1986**	1987*	1988**	Mean	
Canal water	5.40	5.62	6.55	5.86	
Drainage water dilluted, EC$_{IW}$, dS/m					
6	5.36	5.27	6.18	5.60	4.44
9	5.48	4.78	5.70	5.32	9.22
12	5.28	4.22	5.29	4.93	15.87
Drainage water	5.38	3.46	4.98	4.61	21.33
LSD at P = 0.05	NS	0.26	0.35		

Source: Sharma *et al.* (1989).

* Monsoon rains below normal.
** Monsoon rains normal.

d) Disposing of the saline drainage water by injecting it into some appropriate deep aquifer. But care must be taken that it does not affect the quality of groundwater at some other place.

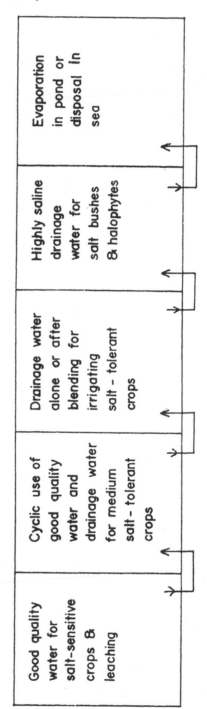

Fig. 8.9 Strategies for reuse of drainage water (Chhabra, unpublished)

e) Drying the highly saline drainage water in evaporation ponds and disposing of the dried salts. However, high water table of the area, high permeability of the soil, and even non-availability of space for digging an evaporation pond due to small land holding, may prevent the use of this technique to salvage salts from the drainage water.

8.12. Change or Blend Water Supply

A change of water supply is a simple but drastic solution to the problems caused by salty irrigation waters. Frequently, this may not be practicable. However, where different sources of water supply are available, a blend may help reduce the hazard of one water. Blending refers to mixing saline with non-saline waters to obtain a composite water supply suitable for irrigation. It is generally done in the irrigation channel itself by regulating the supply of two water sources. Blending helps in reducing salinity as well as RSC of irrigation waters. Any change in quality due to blending may be evaluated using the guidelines in Table 8.1. An example is shown in Table 8.5. Dilution, of course, degrades the better water and improves the poorer water. Whether the result

Table 8.5. Comparison of EC_{IW} and adj. SAR for three different qualities of well waters diluted with canal water

Canal water used	Tubewell water A		Well water X		Well water Y	
	EC_{IW}	adj. SAR	EC_{IW}	adj. SAR	EC_{IW}	adj. SAR
none	3.60	37.40	2.08	40.90	4.00	5.50
20% (1:4)	2.93	32.30	1.71	32.10	3.25	4.82
25% (1:3)	2.76	29.90	1.62	28.70	3.06	4.54
33% (1:2)	2.48	27.00	1.47	26.70	2.75	4.23
50% (1:1)	1.92	21.20	1.16	19.60	2.12	3.56
66% (2:1)	1.35	15.00	0.84	13.10	1.48	2.82
75% (3:1)	1.07	11.00	0.69	9.64	1.17	2.43
80% (4:1)	0.90	9.90	0.60	7.77	0.98	2.12
90% (9:1)	0.57	4.70	0.42	4.14	0.61	1.53
95% (19:1)	0.40	2.99	0.32	2.51	0.42	1.15

Source	Characteristics									
	EC_{IW}, dS/m	Ca^{2+} me/l	Mg^{2+} mel	Na^+ me/l	HCO_3^- me/l	Cl^- me/l	SO_4^{2-} me/l	SAR me/l	pHc	SAR
Canal water	0.23	1.36	0.54	0.48	1.84	0.29	0.17	0.49	7.87	0.76
Tubewell A	3.60	2.48	4.04	32.00	4.46	25.09	8.90	17.72	7.29	37.40
Well X	2.08	0.99	1.20	20.46	10.66	6.03	6.01	19.55	7.31	40.90
Well Y	4.00	16.50	13.80	8.60	2.40	35.70	2.60	2.14	6.83	5.50

is acceptable may depend to a great extent upon the specific situation as to water availability, overall basin water management plans, long range salinity management, and many other factors. Salinity of the resulting blend can be calculated from the following relationship:

$$EC_{IW(mix)} = (EC^1_{IW} \times X_1) + (EC^2_{IW} \times X_2) \qquad (21)$$

where X_1 and X_2 are the fractions of the IW used.

Example

From Table 8.5, a blend is made of 75 per cent canal water ($EC_{IW} = 0.23$ dS/m) and 25 per cent tubewell water (A).

Resulting EC of the blended water

$$= (0.23 \times 0.75) + (3.60 \times 0.75)$$

$$= 0.17 + 0.90$$

$$= 1.07 \text{ dS/m}$$

8.13 Management of Sodic Waters

Sodic waters constitute a significant proportion of groundwaters in arid and semi-arid areas. As discussed in the earlier sections, the prolonged use of such waters immobilises soluble calcium and magnesium in the soil by precipitating them as carbonates. Consequently, the concentration of sodium in the soil solution and on the exchange complex increases and leads to the development of alkali or sodic soil conditions with the following harmful effects:

a) Poor physical properties of soil leading to restricted water and air permeability.

b) A thin but hard crust on the surface that: acts as physical barrier for emerging seedlings and thus reduces plant population, reduces infilteration rate and thus prevents entry of irrigation and rain water into the soil, results in poor air permeability and thus retards root growth, and requires more labour for intercultural operations.

c) Increase in soil pH, which reduces availability of plant nutrients like Ca, N, Zn, and Fe.

d) Excessive availability and consequent toxicity of Na, B, Mo, Li, Se, and other elements.

8.13.1. Treatment of Sodic Waters

Adverse effect of sodic waters on physicochemical properties of the soil can be mitigated through the judicious use of calcium-bearing amendments such as gypsum. By virtue of its low cost and ease in handling, gypsum is by far the most suitable amendment for creating a favourable sodium to calcium ratio in sodic waters.

8.13.2. Gypsum Requirement of Sodic Waters

If the RSC of irrigation water is 2.5 or less, it is generally not necessary to add gypsum. However, for every additional 1 me/l RSC to be neutralised, 86 kg agricultural-grade gypsum (70% purity) is needed per irrigation of 7.5 cm depth per ha. The total amount of gypsum to be added will, however, be determined by the quality of water (RSC to be neutralised) and the quantity of water required for irrigation during a growing season or in a year. For example, if the RSC of the tubewell water is 10.5 me/l and is to be used for growing wheat crop which needs about five irrigations, then the gypsum requirement will amount to 86 (kg gypsum/ha) x 8 (RSC to be neutralised) x 5 (No. of irrigations), i.e., about 3.5 t/ha.

8.13.3. Time and Frequency of Gypsum Application

During the first year, gypsum should be added to both the soil and the irrigation water as dictated by the ESP of the soil and the RSC of the irrigation water, respectively. In subsequent years, application of gypsum is needed on the basis of irrigation water only. Since the same RSC water is to be used year after year, application of gypsum has to be repeated either with each irrigation or at the end of the cropping season. For treatment of water, the best way is to keep a bag of gypsum in the irrigation channel, from where it will slowly dissolve and neutralise the RSC of water.

The best time for application of gypsum to the soil is after the harvest of winter (*Rabi*) crop, preferably in May or June, when some rain has occurred. This will considerably improve the soil prior to the rainy season and will increase the leaching efficiency of rain water, most of which otherwise could be lost because of evaporation and run off due to poor soil permeability. If no rain is received during these months, gypsum application should be postponed till good monsoon showers are received.

Gypsum can be applied even in the standing water as it will hasten leaching of salts and the reclamation process. Upon attaining the proper soil moisture conditions, seed bed should be prepared by shallow ploughing of the soil.

9

Grasses and Trees as Alternate Strategies for Management of Salt-affected Soils

Profitable exploitation of salt-affected soils requires, depending upon the nature of problem, availability of good quality irrigation water, use of amendments, provision of drainage, and an intensive land management system. Sometimes because of the presence of brackish subsoil water and absence of a safe outlet for drainage water, it may not be possible to make the soil non-saline by conventional methods of leaching and drainage. Quite often, the cost of these installations and amendments may be beyond the finanicial reach of the farming community. Moreover, many of these areas belong to village communities or state organisations that may not be willing to put these lands under conventional cropping because they are needed for grazing of cattle and other purposes.

For utilisation and reclamation of these lands, a promising alternative to common grain production is to bring them under a system of trees alone or in association with grasses.This atternative is commonly known as the silvipastral system. In addition to improving the environment and maintaining ecological balance, it provides fuel and fodder, which are also basic needs of the people.

9.1. Reclamative Role of Grasses and Trees

Grasses and trees help in reclaiming salt-affected soils through the following processes:

    a) By removing water from the lower layers, they minimise the capillary rise and thus shift the zone of salt accumulation from the surface to the lower layers. This helps in minimising the salt injury to the shallow-rooted crops.

b) Because of their higher and continuous demand for water, as compared to annual crops, trees intercept the seepage from field channels, canals and higher slopes and thus prevent waterlogging and salinisation of the area. In situations with shallow water table, large-scale planting of trees can biodrain the excess water and lower the water table or prevent its further rise to the surface.

c) Forest litter adds large quantities of OM to the surface soil which improves the physicochemical and biological properties of the soils.

d) Trees help in recycling of plant nutrients from the lower to the upper (surface) layers of soil and thus increase their fertility.

e) Due to intense biological activities of roots and decomposition of forest litter, the pCO_2 of the soil increases, which helps in dissolution of native $CaCO_3$ and thus reclamation of alkali soils (Fig. 9.1).

f) Trees and grasses help in conserving the rain water by increasing the infiltration rate of the soil.

g) Trees and grasses conserve the soil by preventing erosion through greater soil binding, acting as windbreaks, minimising the splashing effect of raindrops, and reducing the peak run-off.

h) By absorbing the salts from the soil and irrigation water, they help in biological harvesting of salts and thus lower the salt content of the soil.

The techniques of cultivating grasses and raising trees and suitability of various species depend upon the nature and severity of the soil problem and the climatic conditions.

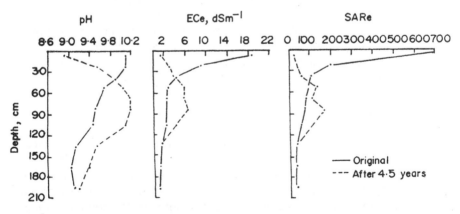

Fig. 9.1.   Changes in soil properties of an alkali soil after 4.5 years of afforestation with *S. aegyptica* (Chhabra *et al.* 1986 a).

9.2. Grasses and Bushes for Salty Soils

9.2.1. Grasses for Alkali Soils

The following grasses are most tolerant and suitable for alkali soils:

Suaeda (*Suaeda martima*)
Karnal grass (*Leptochloa fusca*)
Sporobolus (*Sporobolus marginatus*)
Para grass (*Brachiaria mutica*)
Bermuda grass (*Cynodon dactylon*)
Rhodes grass (*Chloris gayana*)
Gatton panic (*Panicum maximum*)
Blue panic (*Panicum antidotale*)

Among the above, *Leptochloa fusca* commonly known as Karnal grass/Kallar grass is the most tolerant to soil sodicity as well as to waterlogging and holds great potential for areas experiencing heavy monsoon rains. The leaves of this grass are long and linear. It is relatively succulent and palatable. It can be easily propagated by seed or through cuttings. Even under high sodicity conditions, it can give 30 to 40 t of green foliage per hectare without application of fertilisers and amendments. However, with the application of amendments and fertilisers it can yield upto 60 t of fodder per hectare.

Growing of grasses improves both the physical and chemical properties of the soil, largely through intensive biological activity of the roots. It has been observed (Kumar and Abrol, 1984) that after continuous growth of Karnal grass for 3 years the surface soil is improved to so much extent that one can grow a normal crop of rice and wheat without addition of chemical amendments.

Karnal grass can be grown alone or in combination with trees. This grass because of its liking for high pH and sodium grows only in alkali soils and never becomes a weed as it disappears once the soil gets reclaimed.

9.2.2. Grasses and Bushes for Saline Soils

Many salt bushes (*Atriplex* spp.) are tolerant to salinity and desert conditions and have a high protein content (about 25%). Sometimes they yield enough foliage so as to be used as forage crops for livestock and wildlife. One native Californian species produces 25 t of forage per hectare per year. This species can be cultivated, breeded and better selection obtained for increased agricultural production.

Suitable grasses and bushes for inland saline soils are :
Salt cedar (*Tamarix* species)
Suaeda (*Suaeda fruticosa*)

Sporobolus (*Sporobolus pallida*)
Salt bush (*Atriplex* spp.)
Bermuda grass (*Cynodon dactylon*)
Wheat grass (*Agropyron elongatum*)
Eel grass (*Zostera marina*)
Salt grass (*Distichlis spicata*)
Cord grass (*Spartina alternifolia*)
Jaojoba/Hohoba/Goatnut (*Simmondsia chinensis*)

Suitable grasses and bushes for marshy lands are:
Mangroves
Pickleweed (*Salicornia virginica*)
Reed (*Phragmites communistrin*)
Reed (*P. australis*)
Reed (*P. karka*)

Use of rushes and reeds as a means to get rid of excess water, through their high transpiration rate, in the initially highly impermeable soil offers an excellent technique for biodrainage of waterlogged saline soils in temperate regions. It has been reported that reeds can transpire as much as 1,030 mm of water per year (Hemminga and Toorn, 1970). This technology was used to reclaim highly saline soils of Noord-Oost polders in 1942–1948 and Flevopolders in 1956–1972, by the Dutch. After the initial drying of the waterlogged soils through cultivation of reeds, desalinisation through natural drainage becomes a viable proposition. Reeds themselves do not desalinise the soil. They help in early drying of soil through biodrainage, hasten the maturity of the soil and thus improve its leachability. Reeds are also used as a source of energy and natural environment for wildlife in wetlands (Bjork and Graneli, 1978).

9.3. Afforestation

Afforestation of salt-affected soils requires a specialised approach to site development, choice of species, and the level of management. Though many tree species are astonishingly tolerant to adverse conditions, it requires special attention to raise viable plant cover on naturally inhospitable sites. The stress faced by the tree depends upon the nature and extent of the problem in soil, which itself is governed by the soil reaction, nature and amount of soluble salts, physical impediments such as hardpan, quality of underground water, and its depth. Proper understanding the soil problem is a prerequisite for successful plantation of salt-affected soils.

Since the constraints faced by the trees depending upon the type of soil problem i.e., sodicity or salinity, the planting techniques and choice of species also differ in these two situations.

9.3.1. Afforestation of Alkali Soils

Trees, which are essentially deep-rooted plants, experience the following constraints while growing in alkali soils:

a)   High toxicity of Na^+, HCO_3^-, and CO_3^{2-}.
b)   Poor physical conditions due to dispersion of soil particles leading to low water and air permeability. Surface waterlogging during rainy season is a common feature of these soils.
c)   Presence of hard $CaCO_3$ impervious pan commonly known as "Kankar" layer, which acts as a physical impediment for vertical proliferation of the roots.
d)   Poor surface horizontal growth of roots due to high pH and ESP.
e)   Deficiency of plant nutrients such as Ca, N, Fe, and Zn.

Death of growing tips, a typical symptom of die-back disease, necrosis of the leaf tips and margins are quite common symptoms manifested by most tree species leading to their stunted growth and complete failure when raised in alkali soils.

Successful afforestation of alkali soils requires modification in the root environment by (a) amending the chemical nature of soil for optimum growth of roots, leaching of salts, and maximum retention of soil moisture, (b) breaking of hardpan by perforation so as to permit vertical growth of roots, and (c) proper maintenance of soil fertility through application of fertilisers and manures.

a) Site preparation techniques

Since the site preparation is the costliest and most time-consuming operation, considerable planning is required for optimising it. The technique to be adopted for afforestation of alkali soils is mainly governed by the site and soil conditions, species to be planted, and purpose of plantation.

Out of six commonly used methods (Fig. 9.2), the following four are suitable for afforestation of alkali soils:

i)   *Pit method:* Pits of $90 \times 90 \times 90$ cm, $60 \times 60 \times 60$ cm, or $45 \times 45 \times 45$ cm, depending upon the type of species selected for plantation, are most commonly dug for planting trees in normal as well as in alkali soils. However, in alkali soils, when this method is adopted, roots are confined to the pit only. Very few roots cross the underlying hard $CaCO_3$ layer and thus the plants suffer at later stages. Furthermore, depending upon the size of the pit, the requirement for amendment and FYM is generally higher than those in other methods.

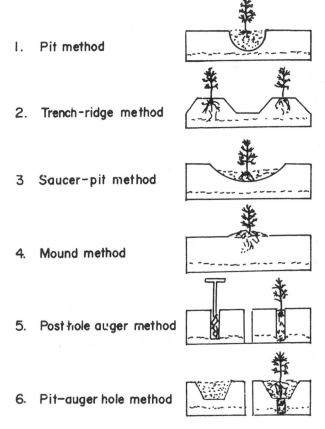

I. Pit method

2. Trench-ridge method

3 Saucer-pit method

4. Mound method

5. Post hole auger method

6. Pit-auger hole method

Fig. 9.2. Methods of planting trees in alkali soils (Chhabra and Kumar, 1989).

ii) *Trench method* : The amendment and FYM are mixed in a trench dug 30 cm deep and 100 cm wide with the help of a tractor. This permits good horizontal proliferation of roots. However, almost all the roots remain in the trench and no roots are found beyond the 30 cm layer. This is an excellent technique for shallow-rooted trees. For big trees with a deep root system it is not a promising technique; wind damage is very likely because of insufficient root development.

iii) *Auger hole method* : In this technique (Abrol and Sandhu, 1980), the hard $CaCO_3$ layer is perforated by drilling a hole with the help of a tractor-mounted auger (Figs. 9.3, 9.4). The auger hole is normally 15 to 20 cm in diameter and 60 to 180 cm in depth. Shallow auger holes, i.e., up to a depth of 60 cm, may be economical but are not successful because the hard layer present below 60 cm depth does not allow the

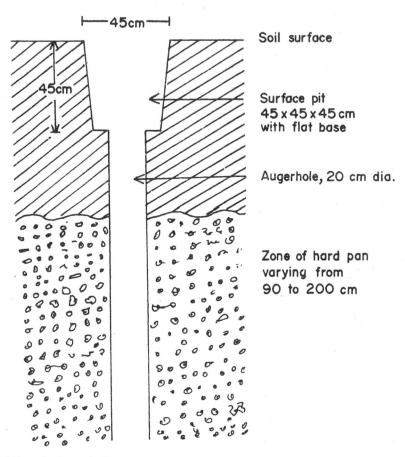

Fig. 9.5.   A schematic diagram showing pit-auger hole method of site preparation for afforestation of highly alkali soils (Chhabra, 1991a).

roots to cross it (Fig. 9.5) and thus prevents the growth of the plant at later stages (Chhabra *et al.*, 1986a).

The auger hole is filled back with amended soil, which provides a favourable environment for the growth of plants. Thus, the roots can bypass the physically and chemically hostile environment of the alkali soils. The roots are able to grow very fast, cross through the perforated hard layer, and reach the wet zone near the water table within 1 to 2 years. Such roots help in meeting the water requirement of the plants from the underground water and thereby decrease the irrigation needs of the plantation. As the volume of soil to be amended in the auger hole is much smaller than that of a pit, the quantity of amendment and manure required is also less, making this technique more economical.

Fig. 9.4.   Root growth of *Sesbania* as affected by depth of auger hole in a highly alkaline soil (a) 180 cm (b) 60 cm (Chhabra and Kumar, 1989).

Fig. 9.3.   Tractor-mounted auger for digging holes for large-scale afforestation of alkaline soils (Chhabra, 1991a).

Fig. 9.6.   Sub-surface channel irrigation method results in accumulation of runoff containing high concentration of alkaline soluble salts and dispersed clay causing death of young plants in alkali soils (Chhabra, unpublished).

iv) *Pit-auger hole method :* In an auger hole system the feeder roots, which confine to the surface layers, are not fully developed. The horizontal growth of these roots is limited to the diameter of the auger hole. As a result, the plants suffer from nutritional deficiencies in the later years of growth. To take care of this, a combination of pit with auger hole in the centre, known as pit-auger hole method (Chhabra and Kumar, 1989), seems to be the best. This technique is now being used by the Forest Departments of the Governments of Haryana and Uttar Pradesh in India for the afforestation of alkali soils under the Social Forestry Programme. Indian Farmers Fertiliser Cooperative Ltd. has also used this technique for afforestation of 2,000 ha of alkali soils in Uttar Pradesh, India.

However, while making auger holes, all the dug-out soil must be taken out with the help of the mechanical auger and not left behind, necessitating reopening before planting (Chhabra, 1991a). This is achieved by:

— making the base of the pit rectangular rather than oval so that it can hold the dug out soil at the surface (Fig. 9.5),

— stopping the circular motion of the auger before taking it out so that the soil stuck in the spiral portion of the auger may come out, and

— slightly reducing the diameter of the cutting edge as compared to the spiral portion of the auger so that all the loosened soil may get stuck in it and could be taken out.

If tilted or angular holes are drilled the roots of trees planted in such holes get stuck and cause abnormal growth of the plants. To avoid this, the auger should be placed in a vertical position before the hole is drilled.

For successful plantation of trees for timber production and orchard establishment, pits of $90 \times 90 \times 90$ cm should be used. In these pits, the hard $CaCO_3$ layer should be perforated with the help of a mechanical auger to a depth of 180 cm. The diameter of the auger hole may vary from 15 to 20 cm. Shallow pits of $60 \times 60 \times 60$ cm or $45 \times 45 \times 45$ cm can be used for energy plantation, especially where *Prosopis juliflora* is to be planted. However, the hard layer must be broken with mechanical auger for proper growth of all other species.

b) Method of refilling pits/auger holes and composition of filling mixture

After digging pits/auger holes, these should be filled back with appropriate mixture as soon as possible. This is necessary for the following reasons :

(i) To avoid accumulation of salts on the interior walls due to evaporation from the exposed surface and physical addition through loose salt-rich surface soil by wind.

(ii) Rain water may fill the pits/auger holes with salt-rich surface runoff from the adjoining areas and thus increase the concentration of toxic soluble salts such as Na_2CO_3 and $NaHCO_3$. If the pits get filled with runoff water, then they should be drained and the fine clay deposited inside should be scraped out before refilling.

(iii) After rains it is difficult to mix amendment, FYM and rice husk (RH) with sticky alkali soil and to properly fill the pits/auger holes.

(iv) If the pits are filled before the rainy season and properly bunded, then the rain water can be used for dissolution of amendment and leaching of salts from the pit.

While the auger holes are being refilled, soil should be properly packed with the help of a wooden stick so as to avoid serious settling down at later stages, which can affect the establishment of seedlings. After filling, a raised earthen bund (ring) should be made around each pit (with the amended soil) so as to prevent the entry of salt-rich runoff water into it.

Composition of filling mixture has a pronounced effect on the survival and growth of the tree species. On the basis of extensive research (Sandhu and Abrol, 1981; Gill and Abrol, 1985; Chhabra and Abrol, 1986; Singh and Abrol, 1986; Chhabra and Kumar, 1989) it has been observed that addition of chemical amendments such as gypsum and pyrites is a must for survival of almost all the tree species raised on alkali soils (Table 9.1). Addition of FYM improves the physical environment as well as supplies nutrients such as N, Fe, Mn, and Zn. Application of physical amendments such as RH, riverbed silt, and sand, improve the leaching of salts and aeration, which are limiting factors in alkali soils. These should be added whereever these are easily and economically available and the texture of the soil is heavy.

The following filling mixtures are suitable for various purposes (Chhabra and Kumar, 1989):

Table 9.1.   Effect of filling mixtures on the performance of 40 months old tree species in a highly alkali soil

Filling mixture composition	*Eucalyptus tereticornis*			*Acacia nilotica*		
	Survival, %	Ht, m	Girth*, cm	Survival, %	Ht, m	Girth*, cm
Original soil (OS)	0	-	-	6	5.84	15.0
OS + 3 kg gypsum	32	4.31	16.8	94	5.48	29.5
OS + 6 kg gypsum	66	5.34	22.6	100	5.49	31.2
OS + 3 kg gypsum + 8 kg FYM	81	6.85	36.2	88	5.91	35.4
Sand + 8 kg FYM + 3 kg gypsum	88	7.94	36.6	100	5.81	38.9
LSD at P = 0.05	40	1.69	9.6	22	NS	NS

*Circumference of stem at 1 m above ground level.
Source: Gill and Abrol (1985).

(i) Shallow pits/auger holes for energy plantation: 3 kg gypsum, 3 kg FYM, and 3 kg RH.

(ii) Deep pits/auger holes for timber production: 5 kg gypsum, 5 kg FYM, and 5 kg RH.

(iii) Deep pits for orchard plantation: 10 kg gypsum, 10 kg FYM, and 5 kg RH.

Replacement of original bad quality alkali soil of the pit/auger hole with good quality normal soil (sometimes referred to as sweet soil) from adjoining fields has been proved beneficial and can be adopted whereever economical (Table 9.2). The filling mixture should be prepared in bulk for a given number of pits/auger holes by mixing the required quantity of amendment, FYM, fertilisers, and the original alkali soil. For this, the top 15 cm soil, which is rich in nutrients, should be used.

Table 9.2. Effect of composition of the filling mixture on the survival and performance of *Albizia* species after 8 years of plantation

Auger hole filling mixture	A. lebbeck		A. falcataria	
	Survival, %	Girth at BH, cm	Survival, %	Girth at BH, cm
Original soil (OS)	0	0	20	3.77
Good soil	100	48.40	100	58.21
OS + 2 kg G	100	46.52	100	34.64
OS + 2 kg G + 8 kg FYM	100	46.55	100	42.05
OS + 2 kg G + 8 kg FYM + RH	100	47.77	100	49.09
LSD at P = 0.05	-	4.00	-	11.84

*BH: Breast height
Source: Chhabra and Abrol (1986).

The quantity of amendment to be added depends upon the volume of the soil, i.e., volume of the auger hole ($\pi.r^2.h$) or pit, soil pH, and the choice of tree species. Orchard plants are more sensitive than forest plants and hence require a higher amount of amendment. Similarly, gypsum requirement for exotic species is greater than that for the local species.

c) Choice of species

Under natural conditions when no amendments are used, the tree species vary greatly in their capacity to tolerate high soil pH/ESP (Table 9.3). Yadav and Singh (1970) conducted a detailed investigation in Vrij Bhumi Forest Division of Uttar Pradesh, India and observed that all the planted species failed to grow on the soils that had pH above 10 and soluble salts above 3.42 per cent in the

Table 9.3.   Relative tolerance of tree species to alkali soil conditions

Alkali condition	pH₂/ESP range	Tree species
High	9.5–10.0 (ESP 25–50)	Mesquite (*Prosopis juliflora*) Acacia (*Acacia nilotica*) Casuarina (*Casuarina equisetifolia*)
Medium	9.0–9.5 (ESP 15–25)	Eucalyptus (*Eucalyptus tereticornis*) Albizia (*Albizia lebbeck*) Albizia (*Albizia falcataria*) Pongamia (*Pongamia pinnata*) Arjun (*Terminalia arjuna*) Sesbania (*Sesbania aegyptica*) Zizyphus (*Zizyphus jujuba*)
Low	8.5–9.0 (ESP < 15)	Parkinsonia (*Parkinsonia aculeata*) Azadirachta (*Azadirachta exotica*) Neem (*Azadirachta indica*) Tamarind (*Tamarindus indica*) Shisham (*Dalbergia sissoo*) Mulberry (*Morus alba*)

top 15 cm soil. *Prosopis juliflora* could tolerate pH up to 10 and soluble salts up to 1 per cent. *Acacia nilotica, Butea monosperma, Dalbergia sissoo, Pongamia pinnata and Terminalia arjuna* were found to grow on soils that were almost free of soluble salts but had pH values upto 9.8. *Allanthus excelsa* appeared to be the least tolerant.

However, when raised after application of amendment and appropriate site preparation techniques, most of the tree species can be successfully raised in alkali soils. On the basis of relative tolerance, local agroclimatic conditions, site preparation, and the purpose of planting, the following species are preferred (Chhabra and Kumar, 1989):

(i) *Energy plantation:* Energy plantation requires those species which can tolerate adverse soil conditions, have higher growth rate, can utilise atmospheric nitrogen through symbiosis, are tolerant to drought, respond easily to coppicing, improve soil conditions through intense biological activities of their roots, provide N-rich litter, and do not require special nursery preparation. Mesquite (*Prosopis juliflora*), Acacia (*Acacia nilotica*), sesbania (*Sesbania aegyptica*), casuarina (*Casuarina equisetifolia*), eucalyptus (*Eucalyptus hybrid* or *E. camaldulensis*), and leucaena (*Leucane leucocephala*) have been found suitable.

(ii) *Timber and minor products:* Acacia (*Acacia nilotica*), casuarina (*C. equisetifolia*), albizia (*A. lebbeck* and *A. falcataria*), eucalyptus (*E. hybrid* and *E. camaldulensis*), arjun (*Terminalia arjuna*), neem (*Azadirachta indica*), jamun (*Eugenia jembolena*), shisham (*Dalbergia sissoo*), khair (*Acacia catechu*), jand (*Prosopis cineraria*), karonda (*Carissa carandes*), tamarind (*Tamarindus indica*), kanzi, and mahua are suitable.

iii) *Orchard plantation:* Depending upon the level of management, agroclimatic conditions, and social needs, amla (*Emlica officinalis*), ber (*Zizyphus jujuba*), Jack-fruit (*Atrocarpus integrifolia*), phalsa (*Grevia subinae*), and bail (*Aegle marmelos*) may be considered. For ber, plantation should be done with a wild variety and, after establishment of the rootstock, budding with good quality scion should be done *in situ*. Mango and guava are sensitive and should not be planted in alkali soils. Bamboo and date-palm can also be planted, preferably on roadsides and boundaries.

iv) *Grasses:* Since community lands are also used for grazing cattle, it is desirable, that along with trees, suitable grasses should also be raised on these lands. In addition to providing much needed fodder (Table 9.4), these grasses hasten the reclamation process (Table 9.5) through the activities of their roots and biological harvesting of salts. For alkali soils Karnal grass (*Leptochloa fusca*) is most tolerant. It can tolerate not only high pH/ESP but also the waterlogging that is common in these soils. It can be propagated by cuttings as well as seed. Other promising grasses for alkali soils are Bermuda

Table 9.4.   Green forage yield of Karnal grass grown in association with *Prosopis juliflora*

Planting year	Number of cuttings per year	Cutting month	Forage yield, t/ha
1	1	November	2.2
2	4	May to October	13.1
3	3	July to September	10.1
4	3	May to September	7.6
5	4	July to October	13.5
Total	15		46.5

Source: Chhabra and Singh (1990).

Table 9.5.   Properties of surface 15 cm of an alkali soil as affected by *Prosopis juliflora–Leptochloa fusca* agroforestry system

Soil property	Without grass			With grass		
	0	22	52	0	22	52
	Months after planting					
pH	10.30	10.00	9.70	10.30	9.70	9.40
EC, dS/m	2.20	0.83	0.66	2.20	0.70	0.42
OC, %	0.18	0.20	0.30	0.19	0.28	0.43
Available N, kg/ha	79.00	83.00	100.00	82.00	96.00	139.00
Available P, kg/ha	35.00	33.00	30.00	35.00	29.00	22.00
Available K, kg/ha	543.00	519.00	528.00	543.00	471.00	402.00

Source: Chhabra and Singh (1990).

grass (*Cynodon dactylan*), para grass (*Brachiaria mutica*), Rhodes grass (*Chloris gauana*), and blue panic (*Panicum antidotale*).

d) Methods of planting

Normally, seedlings raised in poly-pots in nursery are used for planting in the field. However, *Sesbania* seeds can be sown directly in the pit. For this, after irrigation, 8 to 10 seeds should be put on the surface of the pit and then covered with rice husk or straw, which acts as a mulch. It is better to sow presoaked seeds and to keep the surface moist by frequent irrigations. After germination, 2 to 3 plants per pit can be retained for energy plantation. *Prosopis juliflora* and *A. nilotica* can also be raised directly from seeds sown in gypsum-amended trenches or ridges.

e) Time of planting

For most areas, saplings should be planted after the onset of rains (July–September). It is advisable that the first two or three rain showers should be allowed to flush out salts accumulated on the surface, to leach salts of the soil in the pit/auger hole, and to let the soil in the pit to settle down. This provides salt-free and amended environment for the better establishment of young plants. Once established, the warm and humid climate during this period helps in rapid growth of the young sapling. Deciduous trees like shisham and poplar should be planted in the spring (February–April). Trees planted in the spring need regular irrigation until the rainy season because the climate during April to June is very hot and dry. This, in addition to increasing the water demand of the growing plant, results in higher accumulation of salts in the root environment. To mitigate the effect of harsh temperature and high salts, frequent irrigations with good quality water are required. Normally, survival of the saplings planted in spring is low when assured irrigation is lacking.

f) Irrigation

Irrigation during initial stages of establishment and growth is essential for trees raised on alkali soils (Table 9.6). As the distribution of rainfall even during the

Table 9.6.   Effect of irrigation on survival, height, DBH and of *Prosopis juliflora* at 2 years after planting and aerial woody biomass obtained 33 months after planting

Treatment	Survival, %	Height, m	DBH*, cm	Woody biomass, t/ha	
				Lopped	Harvested
Unirrigated	64	1.37	2.12	1.89	3.38
Irrigated	91	2.14	2.78	4.00	7.78
LSD at P = 0.05	18	0.32	0.45	0.92	1.26

*DBH= Diameter at breast height
Source: Chhabra and Singh (1990).

rainy season is uneven, there may be periods of prolonged drought neces-sitating irrigation in the rainy season also. Depending upon the climatic conditions and distribution and frequency of rainfall, irrigation should be applied at least once in 7 days in the first 3 months and then once in a month for 1 year. Even at the later stages protective irrigation is required for most of the plantations raised in alkali soils. Since root growth is limited, frequency of irrigation should be greater for plants raised in shallow pits without auger holes.

As the surface soil contains high concentration of salts, irrigation through flood method or sub-surface irrigation channels (Fig. 9.6) should be avoided in the first year of planting. Spot irrigation with the help of containers, plastic pipes or dripper is more useful. At later stages, irrigation can be given through channels made above the ground and joining the various pits. However, to prevent salts from accumulating on the ridges of the channels, reaching the pits, and causing secondary sodication, application of gypsum in the channels and on the ridges is of great help. Physical removal of salts (wherever possible), especially in the first year, can be useful.

g) Drainage
Most of the trees are sensitive to excess water in the rootzone. During the rainy season, alkali soils because of their low elevation and poor infiltration rate experience floods. This excess water can cause temporary waterlogging and kill the young plants. To avoid this, excess rain water should be removed, through surfaces drains, as early as possible.

However, efforts should be made to retain rain water, as much as possible, within the field itself but away from the active rootzone. This can be done by making earthen rings around the pits and bunds around the field. These bunds will also prevent entry of runoff water from outside. Conservation of rain water in small flat plots can be useful as the submergence caused by it helps in ameliorating the alkali soils by enhancing the dissolution of native $CaCO_3$.

h) Fertilisation
Alkali soils are rich in available phosphorus and potassium (Chhabra 1985). Chhabra (unpublished) further observed that height and girth of 8-year-old eucalyptus raised in a highly alkali soil did not show any significant difference as a result of P and K application (Table 9.7). Hence there is no need to apply these nutrients at the time of planting. Nitrogen is the most limiting nutrient and thus trees like eucalyptus and other plants which cannot avail atmospheric N through symbiosis with *Rhizobium* or *Frankia* require chemical fertilisers. Nitrogen at a rate of 25 g per plant mixed in the filling mixture and its regular application every year (Fig. 9.7) has proved beneficial for eucalyptus (Chhabra and Singh, 1990). Since the development of roots in the auger hole is restricted and uptake of nutrient by tree species is much slower, application of nitrogen

Table 9.7. Effect of different levels of P and K on the growth of eucalyptus in a highly alkali soil

Treatment		Height, m	Girth at BH, cm
P_0	K_0	15.80	39.85
P_{13}	K_0	15.90	38.66
P_{26}	K_0	16.32	38.97
P_0	K_{25}	17.51	40.23
P_0	K_{50}	16.80	38.03
P_{13}	K_{25}	16.00	39.60
LSD at P = 0.05	-	NS	NS

* BH = Breast height,
P_{13} = 30 g P_2O_5 or 190 g SSP/plant.
K_{25} = 30 g K_2O, or, 50 g MP /plant.
Filling mixture : OS + RH + FYM + gypsum.
N = 110 g/plant/year.
Source: Chhabra (unpublished).

Table 9.8. Effect of auger hole mixture and nutrients on the growth of eucalyptus in a highly alkali soil after 8 years of plantation

Treatments	Girth at BH, cm
OS	
OS + 2 kg gypsum (G)	24.52
OS + RH + 2 kg G	28.92
OS + RH + 50 g N + 2 kg G	40.49
OS + RH + 100 g N + 2 kg G	40.06
OS + RH + 200 g N + 2 kg G	41.06
OS + RH + 100 g N + 10 g $ZnSO_4$ + 2 kg G	44.97
OS + RH + 10 kg FYM + 2 kg G	40.66
OS + RH + 100 g N + 10 g $ZnSO_4$ + 10 kg FYM + 2 kg G	42.31
LSD at P = 0.05	4.40

*BH = Breast height
Source: Chhabra (unpublished).

at a higher rate may damage the plants. Hence only lower doses of N, i.e., 25 to 50 g per plant and that too with assured irrigation should be provided. However, higher dose of N, i.e., 100 to 200 g/plant/year, should be applied after 1 year of growth to meet the increased requirement (Table 9.8). Fertilisers should be applied in April and September, when fresh growth is taking place.

To prevent Zn deficiency, 10 g of $ZnSO_4$ per plant should be applied in the filling mixture. Sometimes iron chlorosis due to high pH and $CaCO_3$ induced iron unavailability has been observed in eucalyptus grown in alkali soils.

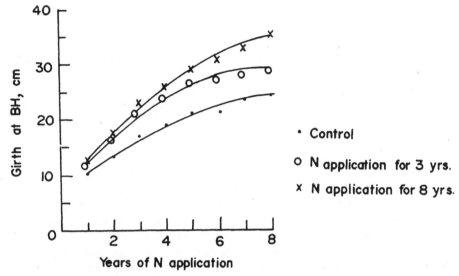

Fig. 9.7. Effect of N application, 25 g/tree, on the girth of eucalyptus raised in a highly alkali soil (Chhabra and Singh, 1990).

Application of pyrites and FYM does help in meeting the iron needs of the trees.

i) Spacing and pruning

It is an established fact that with increase in number of plants per unit area, there is an increase in total biomass production, although the biomass per tree, decreases. So, for energy plantation, the distance can be as title as 1 m from plant to plant and row to row. Since the overall growth of trees in alkali soils is less than that in normal soils, it is observed (Table 9.9) that planting distance can be reduced from 3 × 3 m to 1.5 × 1.5 m in case of eucalyptus to obtain

Table 9.9. Effect of spacing on wood production by 8-year-old plantation of eucalyptus raised in an alkali soil

Spacing, m	No. of plants per ha	Sur., %	Effective no. of plants/ha after 8 years	Girth at BH, cm	Height, m	Vol. of wood, m³/ha	Weight of dry wood, t/ha
3.0 × 3.0	1089	90	980	40.49	15.95	43.20	34.56
3.0 × 1.5	2178	90	1960	36.57	16.20	71.70	57.36
1.5 × 1.5	4356	90	3920	35.20	16.70	146.77	117.42
LSD at P = 0.05		NS		1.30	0.60	10.40	9.60

BH = Breast height
Source: Chhabra (unpublished).

higher wood production. In alkali soils, closer planting aiso partly covers the risk of high mortality in the initial stages. For timber production, the distance can be regulated later by selective felling.

Except for eucalyptus, stem (about 1/3 of height) should be pruned annually to remove unwanted branches. This helps accelerate the growth of the tree, improves the quality of wood, and also promotes the growth of understorey vegetation. In the case of *P. juliflora*, selective lopping helps in proper training and timber production. Pruning should be done before the onset of spring, i.e., in January or February.

j) Mixed plantation
Considering the hostile soil conditions, mortality of young saplings of many species in the initial stages may be quite considerable. Some species grow much slower and require a long time for economic exploitation. In view of this and to avoid complete failure, instead of monoculture, planting should be done with mixed species. Since *Prosopis juliflora* is quite tolerant and can provide good cover in the initial years, 50 to 70 per cent of the planting should be done with it and the rest with other species. The distance between the rows and the level of site preparation should be so adjusted that the big plants may be in bigger pits and at wider spacing while plants such as mesquite and sesbania are planted in small pits and at closer spacing. Later on lower-storey plants can be harvested for fuel wood production, while others can continue to grow for timber or fruit production. However, while mixing species their growing habits, especially the aggressive index, should be kept in mind.

k) Cost of plantation
The major expenditure on planting trees in alkali soils is on site development, which involves land shaping and digging of auger holes, cost of amendment, and providing higher number of irrigations. This is in addition to the expenditure on other operations such as cost of plants and providing cattle protection trench.

9.3.2. Afforestation of Saline Soils

The tree growth in saline soils is affected by high osmotic pressure of soluble salts resulting in low water availability, toxic effects of soluble ions such as Na^+, Cl^-, SO_4^{2-}, and low availability of nutrients.

In arid saline soils, the plants face stress due to excessive accumulation of salts in the rootzone, while in waterlogged saline soils or saline soils with shallow water table, salinity, poor aeration, and excess moisture affect simultaneously and do not allow them to establish and grow normally. For such soils, care has to be taken to prevent excessive salt accumulation and also to provide aeration in the rootzone.

Various aspects of raising trees in saline soils are discussed below:

a) Planting methods

i)*Ridge-trench method:* In this method trees are planted on ridges, 50 to 100 cm high, and trenches between the ridges are used for draining excess water. In coastal saline soils where because of ingress of sea water the soils become saline or because of shallow water table the salinity of the surface soil increases, the planting of trees on ridges is useful. Rainfall is normally high near the sea-coast, and this helps in leaching of salts from ridges and thus providing stress-free environment. This also improves aeration in the rootzone. However, in the plains and especially under arid conditions, the salts have a tendency to accumulate on the surface (Fig. 9.8a; Tomar and Gupta, 1986). Under such conditions, because of greater evaporation from the exposed area and little leaching during monsoons, excess salts accumulate on the sides and top of the ridges. Furthermore, as a result of greater evaporation, moisture content is also generally lower in the ridges than in the original land configuration. Thus, in this method plant roots face excess salts and low moisture conditions, making it hard for them to survive.

ii*) Sub-surface planting in auger holes:* Since, in saline soils, the maximum salt accumulation takes place on the surface, greater success can be achieved if the young plant roots are placed in the sub-surface, which contains less salts. For this, planting saplings in auger holes 30–45 cm below the surface helps them to avoid the zone of excessive salt accumulation and the roots are exposed to only the intermediate salinity, which corresponds to the transmission zone salinity of the soil (Fig. 9.8b).

However, the auger holes often collapse because runoff from the adjoining areas and the salts accumulate in these places. Furthermore, though sub-surface salinity is much lower than the surface salinity, it may be high enough to kill the plants in the initial critical stage of establishment.

iii*) Planting in furrow-cum-irrigation channels:* Young saplings can be planted in shallow trenches or furrows, about 30 cm deep, which are also used for irrigation (Fig. 9.8c). The water in the furrow pushes the salts away from the active rootzone and also serves as a source of irrigation for the plants. Considerable success has been achieved with this technique at the Experimental Farm of the Central Soil Salinity Research Institute, Karnal, India.

b) Use of amendments

Chemical amendments such as gypsum or pyrites are not needed for raising plantation in saline soils. However, in the case of heavy-textured soils, addition of sand, silt, or rice husk improves the physical environment and allows more efficient leaching of salts. Application of organic mulch also helps in reducing the accumulation of salts on the surface. In areas having high RSC waters,

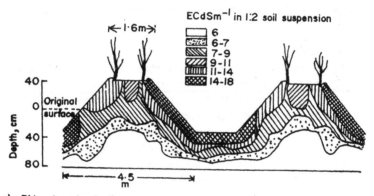

a) Ridge trench planting method

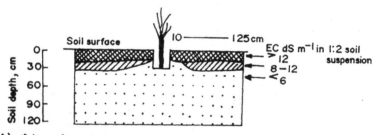

b) Sub-surface planting method

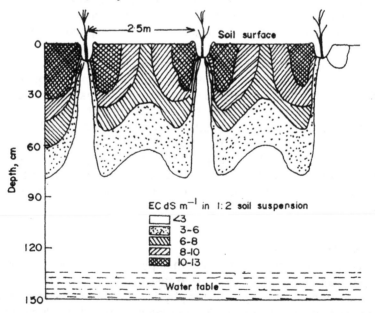

c) Channel planting method

Fig. 9.8.   Zone of salt accumulation and methods of tree plantation in saline soils (Tomar and Gupta, 1986).

regular use of gypsum to counteract the harmful effect of CO_3^{2-} and HCO_3^- of irrigation water is beneficial.

c) Irrigation

Regular irrigation to trees especially in the initial period of establishment is essential for afforestation of saline soils. Although the soil may look moist and have enough water in many areas because of shallow water table, irrigation with good quality water is a must, mainly to leach down the salts from the rootzone and thereby provide a stress-free environment. For successful establishment of plants, at least 8 to 10 irrigations should be provided in the first year and 3 to 6 in the second year of plantation. Even at later stages, whereever available, protected irrigation should be provided.

d) Choice of species

Barhava and Rawati (1967) observed that in dry $Cl^-SO_4^{2-}$ solonchaks of Israel (ECe varying from 12 to 17 dS/m), *Eucalyptus camaldulensis* and *Pinus helepensis* were found to be most tolerant. Migunova (1976) reported that out

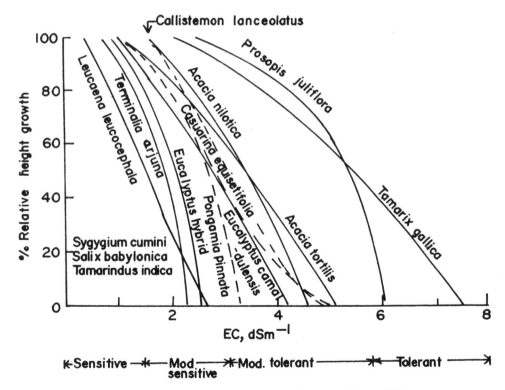

Fig. 9.9. Relative tolerance of tree species to salinity (Tomar and Gupta, 1986).

of several species *Tamrix ramosissima* and *Terminalia tetranda* could root successfully in the horizons that contained up to 6 per cent salts. Bangash (1977) in a trial with seeds of *Acacia nilotica, Albizia lebbeck, Parkinsonia aculeata, Prosopis spicigera*, and *Zizyphus jujuba* observed that *A. nilotica, A. lebbeck* and *Z. jujuba* were tolerant to NaCl salinity conditions.

Tree species differ in their ability to withstand salinity and aeration stress individually and simultaneously. In areas where salinity is not associated with shallow water table conditions, species such as *Acacia auriculiformis, Terminalia arjuna*, and *Leucaana leucocephala* can be grown. Where high salinity and shallow water table exist together, *Casuarina equisetifolia, Tamrix articulata*, and *P. juliflora* have been found to perform well. For moderate salinity, 10–22 dS/m Tomar and Gupta (1986) reported that *A. nilotica, A. tortilis, A. auriculiformis, C. equisetifolia*, and *E. camaldulensis* can be grown successfully (Fig. 9.9).

References

Abrol, I.P., and Acharya, C.L. 1975. Soil water behaviour and irrigation frequencies in soils with physical constraints. Proc. Second World Cong. Intern. Water Resour. Assoc., New Delhi, India. 1: 335–342.

Abrol, I.P., and Bhumbla, D.R. 1971. Saline and alkali soils in India—Their occurrence and management. FAO World Soil Resources. Rep. No. 41: 42–51.

Abrol, I.P., and Bhumbla, D.R. 1979. Crop response to differential gypsum application in a highly sodic soil and tolerance of several crops to exchangeable sodium under field conditions. Soil Sci. 127: 79–85.

Abrol, I.P., Chhabra, R., and Gupta, R.K. 1980. A fresh look at the diagnostic criteria for sodic soils. Proc. Intern. Symp. Salt-affected Soils, Karnal, India. pp. 142–147.

Abrol, I.P., and Dahiya, I.S. 1974. Flow associated precipitation reaction in saline-sodic soils and their significance. Geoderma. 11: 305–312.

Abrol, I.P., Dahiya, I.S., and Bhumbla, D.R. 1975. On the method of determining gypsum requirement of soils. Soil Sci. 120: 30–36.

Abrol, I.P., Gupta, R.K., and Singh, S.B. 1979. Note on solubility of gypsum and sodic soils reclamation. J. Indian Soc. Soil Sci. 26: 98–105.

Abrol, I.P., and Sandhu, S.S. 1980. Growing trees in alkali soils. Indian Fmg. 30(6): 19–20.

Abrol, I.P., and Verma, K.S. 1978. A comparative study of the effect of gypsum and pyrites on yield of rice and wheat grown in a highly sodic soil. Proc. FAI-PPCL Sem. Use of Sedimentary Pyrites in Reclamation of Alkali Soils. Lucknow, India. pp.136a–136c.

Acharya, C.L., Sandhu, S.S., and Abrol, I.P. 1979. Effect of exchangeable sodium on the rate and pattern of water uptake by raya (*Brassica juncia*) in the field. Agron. J. 71: 936–941.

Agarwal, A.S., Singh, B.R., and Kanehiro, Y. 1971. Ionic effect of salts on mineral nitrogen release in an Alophanic soil. Soil Sci. Soc. Am. Proc. 35: 454–457.

Alperovitch, N., Shainberg, I., and Keren, R. 1981. Specific effect of magnesium on the hydraulic conductivity of sodic soils. J. Soil Sci. 32: 543–554.

Al-Rawi, A.H., and Sadallah, A.M. 1980. Effect of urea and salinity on the growth and yield of wheat. Proc. Intern. Symp. Salt-affected Soils. Central Soil Salinity Research Institute, Karnal, India. pp. 433–439.

Amer, F., El-Elgabaly, M.M., and Balba, M.A. 1964. Cotton response to fertilization of two soils differing in salinity. Agron. J. 56: 208–211.

Antipov-Karataev, I.M., and Manaeva, L.Y. 1958. The role of magnesium in solonetz properties. Agrokm. es. Talajt. 7: 1–14.

Anonymous. 1973. Water Quality Criteria-1972. Ecological Research Series. ESA-3/73/003 U.S. Environ. Protec. Agency (EPA), Washington, 594.

Arnay, S. 1956. Contribution to the role of magnesium in the formation of alkali soils. Proc. Sixth Intern. Cong. Soil Sci. 11: 655–659.

Arulanandan, K., and Heinzen, R.T. 1977. Factors influencing erosion in dispersive clays and methods of identification. Proc. Sem. Erosion and Solid Matter Transport in Inland Waters, Paris, IAHS-AISH Pub. 122: 75–85.

Ashok Kumar., and Abrol, I.P. 1982. Note on the effect of gypsum levels on the boron content of soils and its uptake by five forage grasses in a highly sodic soil. Indian J. Agric. Sci. 52: 615–617.

Ayers, A.D., and Wadleigh, C.H. 1960. Saline and sodic soils in Spain. Soil Sci. 90: 133–138.

Ayers, R.S., and Westcot, D.W. 1976. Water quality for agriculture. Irrigation and Drainage Paper 29, FAO, Rome. p. 97.

Ayers, R.S., and Westcot, D.W. 1985. Water quality for agriculture. Irrigation and Drainage Paper 29, FAO, Rome. Rev. 1: 174.

Ayoub, A.T. 1960. Preliminary study of salinity problems in northern Sudan. FAO Symp., Baghdad.

Baddesha, H.S., and Chhabra, R. 1985. Sewage utilization through forestry. Proc. Natn. Sem. Pollution and Environment. NEERI, Nagpur, India.

Baddesha, H.S., Rao, D.L.N., Abrol, I.P., and Chhabra, R. 1986. Irrigation and nutrient potential of raw sewage waters of Haryana, India. J. Agric. Sci. 56: 584–591.

Bains, S.S., and Fireman, M. 1968. Growth of sorghum in two soils as affected by salinity, alkalinity and phosphorus levels. Indian J. Agron. 13: 103–111.

Balba, A.M. 1960. Effect of sodium water and gypsum increments on soil chemical properties and plant growth. Alexandria J. Agric. Res. 8: 51–54.

Balba, A.M. 1980a. (Soil Fertility and Fertilization) Dar El-matbouat Al-Gadidah (in Arabic).

Balba, A.M. 1980b. Minimum Management Programme to combat World Desertification. UNDP Consultancy Rep. Adv. Soil Water Res., Alexandria.

Balba, A.M., and Soliman, M.F. 1978. Quantitative evaluation of the relative specific effect of cations on the plant growth and nutrient absorption. Proc. 11th Intern. Soil Sci. Soc. Cong. Edmonton. 1: 319–320.

Bandyopadhya, A.K. 1974. Seasonal variation of boron in saline sodic soils. Ann. Arid and Semi-arid Zones. 13: 125–128.

Bangash, S.H. 1977. Salt tolerance of forest tree species as determined by germination of seeds at different salinity. Pakistan J. For. 27: 93–97.

Barhava, N.B., and Rawati, B. 1967. The tolerance of some species of eucalyptus, pines and other forest trees to soil salinity and low soil moisture in the Negev. Israel J. Agric. Res. 17: 65–76.

Beek, C.G., and van Breemen, N. 1973. The alkalinity of alkali soils. J. Soil Sci. 24: 129–136.

Ben-Zioni, A., Itai, C., and Vaadia, Y. 1967. Water and salt stresses : Kinetin and protein synthesis in tobacco leaves. Pl. Physiol. 42: 361–365.

Bernstein, L. 1962. Salt-affected soils and plants. Proc. Intern. Symp. Problems of Arid Zones, Paris. UNESCO Pub. pp. 139–149.

Bernstein, L., and Francois, L.E. 1973. Comparison of drip, furrow, and sprinkler irrigation. Soil Sci. 115: 73–86.

Bernstein, L., Francois, L.E., and Clark, R.A. 1974. Interactive effects of salinity and fertility on yield of grains and vegetables. Agron. J. 66: 412–421.

Bernstein, L., and Hayward, H.E. 1962. Physiology of salt tolerance. Ann. Rev. Physiol. 9: 25–46.

Bernstein, L., and Ogata, G. 1966. Effects of salinity on nodulation, nitrogen fixation and growth of soybeans and alfalfa. Agron. J. 58: 201–203.

Bernstein, L., and Pearson, G.A. 1956. Influence of exchangeable sodium on the yield and chemical composition of plants—1. Green beans, garden beets, clover and alfalfa. Soil Sci. 82: 247–258.

Bhardwaj, K.K.R. 1975. Survival and symbiotic characteristics of rhizobium in saline-alkali soils. Pl. Soil. 43: 377–385.

Bhardwaj, K.K.R., and Abrol, I.P. 1978. Nitrogen management in alkali soils. Proc. Natn. Symp. Nitrogen Assimilation Crop Productivity, Hisar, India. pp. 83–96.

Bhargava, G.P., and Abrol, I.P. 1978. Characteristics of some typical salt-affected soils of Uttar Pradesh. Rep. No. 6. Central Soil Salinity Research Institute, Karnal, India.

Bhargava, G.P., Abrol, I.P., and Bhumbla, D.R. 1976. On the nomenclature of some salt-affected soils of the Indo-Gangetic plains. J. Indian Soc. Soil Sci. 24: 81–83.

Bhargava, G.P., Sharma, R.C., Pal, D.K., and Abrol, I.P. 1980. A case study of the distribution and formation of salt-affected soils in Haryana state. Proc. Intern. Symp. Salt-affected Soils, Karnal, India. pp.83–91.

Bhatnagar, R.K., Gupta, V.K., and Mathur, C.M. 1966. Manganese status in medium black soil of Rajasthan with special reference to Chambal command area. J. Indian Soc. Soil Sci. 14: 173–176.

Bhatt, M.N., Goel, K.N., and Uprety, M.C. 1978. Use of pyrites in reclamation of alkali soils in Uttar Pradesh. Proc. FAI-PPCL Sem. Use of Sedimentary Pyrites in Reclamation of Alkali Soils, Lucknow, India. pp. 107–114.

Bhattacharya, R.K. 1972. Fertilizer requirement of rice grown in costal saline soils of Sunderban, West Bengal, India. Indian J. Agron. 17: 69–73.

Bhola, S.N., Dhir, R.P., and Sharma, B.K. 1980. Effect of saline irrigation water on the protein content of wheat. Indian J. Agron. 25: 546–548.

Bhumbla, D.K., Chhabra, R., and Abrol, I.P. 1980. Distribution of boron in alkali soils. Ann. Rep. Central Soil Salinity Research Institute, Karnal, India.

Bhumbla, D.R. 1977. Alkali and saline soils of India. Proc. Indo-Hungarian Sem. Management of Salt-affected Soils, Karnal, India. pp. 14–19.

Bhumbla, D.R., and Abrol, I.P. 1972. Is your water suitable for irrigation. Indian Fmg. 22(4): 15–16.

Bhumbla, D.R., and Chhabra, R. 1982. Chemistry of sodic soils. In Review of Soil Research in India. Part 1. 12th Intern. Cong. Soil Sci., New Delhi, India. pp. 160–180.

Bhumbla, D.R., and Dhingra, D.R. 1964. Micronutrient status of saline and alkali soils of the Punjab. J. Indian Soc. Soil Sci. 12: 255–260.

Biswas, C.R. 1975. Management of nitrogen for rice in waterlogged coastal saline soils. Ann. Rep. Central Soil Salinity Research Institute, Karnal, India.

Bjork, S., and Graneli, W. 1978. Energy reeds (in Swedish, summary in English) CODEN LUNBDS, Limnologiska Institute. Lund University, Lund.

Bower, C.A., and Maasland, M. 1963. Sodium hazards of Punjab groundwaters. Proc. Symp. Waterlogging and Salinity in West Pakistan. pp. 47–61.

Buringh, P., and Edelman, C.H. 1955. Some remarks about soils of the alluvial plains in Iraq, South of Baghdad. The Netherlands J. Agric. Sci. 3: 40–49.

Chauhan, R.K.S. 1978. Note on comparative study of pyrites and gypsum as soil amendments in alkali soils. Proc. FAI-PPCL Sem. Use of Sedimentary Pyrites in Reclamation of Alkali Soils, Lucknow, India. pp. 127–129.

Chawla, K.L., and Abrol, I.P. 1982. Effect of gypsum fineness on the reclamation of sodic soils. Agric. Water Mgmt.. 5: 41–50.

Chawla, V.K. 1969. Available nitrogen and phosphorus status of saline sodic soils in Punjab, India. Agron. J. 64: 361–362.

Chhabra, R. 1976a. Effect of different levels, sources and methods of zinc application on growth of rice and wheat in sodic soils. Ann. Rep. Central Soil Salinity Research Institute, Karnal, India. pp. 29–30.

Chhabra, R. 1976b. Phosphorus availability in sodic soils. Ann. Rep. Central Soil Salinity Research Institute, Karnal, India. pp. 27–28.

Chhabra, R. 1985. Crop responses to phosphorus and potassium fertilization of a sodic soil. Agron. J.77 : 699–702.

Chhabra, R. 1986. Comparative efficiency of different phosphatic fertilizers in reclaimed alkali soils. Ann. Rep. Central Soil Salinity Research Institute, Karnal, India. p. 27.

Chhabra, R. 1987a. Fertilizer management in sodic soils. National Workshop on Reclamation and Management of Alkali soils. Ministry of Agriculture, New Delhi, India.

Chhabra, R. 1987b. Forestry—a cure for sewage pollution. Aquaworld. 11(6): 185–187.

Chhabra, R. 1988a. Conservation of nutrients and irrigation potential of domestic waste waters (sewage waters) sans pollution. J. Soil Water Conserv. India. 32: 113–123.

Chhabra, R. 1988b. Application of soil testing for salt-affected soils. Fertil. Marketing News. 19(2): 1–7.

Chhabra, R. 1988c. Beware of killer gas in pits of tubewells. Indian Fmg. 38(6): 39–43.

Chhabra, R. 1989. Sewage water utilization through forestry. Bull. No. 15. Central Soil Salinity Research Institute, Karnal, India.

Chhabra, R. 1990. Localised irrigation using ceramic pot pitchers. Siphon News Letter. Center for Irrigation Engineering. Catholic University Leuven, Leuven, Belgium. April, 1990: 15–17.

Chhabra, R. 1991a. Pit-auger hole method—a site preparation technique for afforestation of alkali soils. Forest Department, Government of Haryana, India.

Chhabra, R. 1991b. Feasibility report on control of pollution by using distillery effluent through forestry. Jagatjit Industries Ltd., Hamira, Punjab, India.

Chhabra, R. 1991c. Feasibility report on control of pollution by using paper mill effluent through forestry. Shreyans Paper Mills Ltd. Ahmedgarh, Sangrur, Punjab, India.

Chhabra, R., and Abrol, I.P. 1975. Effect of pyrites and phosphorus on the yield and chemical composition of rice in sodic soils. Ann. Rep. Central Soil Salinity Research Institute, Karnal, India. pp. 48–49.

Chhabra, R., and Abrol, I.P. 1977. Reclaiming effect of rice grown in sodic soils. Soil Sci. 124: 49–55.

Chhabra, R., and Abrol, I.P. 1981. Fertilizer phosphorus requirement of rice and wheat grown on sodic soils, Intern. Rice Res. Institute, Philippines, News Letter. 6(4): 22.

Chhabra, R., and Abrol, I.P. 1983. Principles governing fertilization in salt-affected soils. Fertil. Industry Ann. Rev. India. 11: 147–159.

Chhabra, R. and Abrol, I.P. 1986. Effect of amendments and nutrients on the performance of selected tree species in sodic soils. Ann. Rep. Central Soil Salinity Research Institute, Karnal, India. pp. 23–25.

Chhabra, R., and Abrol, I.P. 1987. Fertilizer management in salt-affected soils. Proc. ICAR-IRRI Symp. Soil fertility constraints and fertilizer use strategy in rice based cropping systems. New Delhi, India.

Chhabra, R., Abrol, I.P., and Chawla, K.L. 1986a. Tolerance and productivity of *Sesbania species* for fuel wood production in sodic soils. Proc. Intern. Symp. Afforestation of Salt-Affected Soils. Central Soil Salinity Research Institute, Karnal, India. pp. 30–48.

Chhabra, R., Abrol, I.P., Dargan, K.S., and Gaul, B.L. 1980a. Save on phosphatic fertilizers in the initial years of alkali soils reclamation. Indian Fmg. 30(3) : 13–15.

Chhabra, R., Abrol, I.P., and Singh, M.V. 1980b. Leaching losses of phosphorus in sodic soils. Proc. Intern. Symp. Salt-affected Soils. Central Soil Salinity Research Institute, Karnal, India. pp. 418–422.

Chhabra, R., Abrol, I.P., and Singh, M.V. 1981. Dynamics of phosphorus during reclamation of sodic soils. Soil Sci. 132: 319–324.

Chhabra, R., Flowers, T., and Yoe, J. 1986b. Absorption of P by rice as affected by pH and salinity of nutrient solution. Deputation Rep. Indo-British Cooperation. Central Soil Salinity Research Institute, Karnal, India.

Chhabra, R., and Kumar, V. 1989. Afforestation for development of salt-affected soils. ICAR-IFFCO Pub. Central Soil Salinity Research Institute, Karnal, India. p. 27.

Chhabra, R., Pleysier, J., and Cremers, A. 1975. The measurement of cation exchange capacity and exchangeable cations in soils—A new method. Proc. Intern. Clay Conf. Mexico. pp. 439–449.

Chhabra, R., Rajput, R.K., and Singh, N.T. 1988. Reclamation and utilization of salt-affected soils in India—A status paper. Third Meeting of the Cooperative for Research on Problem Soils, Bagor, Indonesia.

Chhabra, R., Ringoet, A., and Lamberts, D. 1976. Kinetics and interaction of chloride and phosphate absorption by intact tomato plants (*Lycoperscon esculentum* Mill) from a dilute nutrient solution. Z. Pflanzenphysiologie. 78: 253– 264.

Chhabra, R., Ringoet, A., Lamberts, D., and Scheys, I. 1977. Chloride losses from tomato plants (*Lycoperscon esculentum* Mill). Z. Pflanzenphysiologie. 81: 89–94.

Chhabra, R., Singh, A., and Abrol, I.P. 1980c. Fluorine in sodic soil. Soil Sci. Soc. Am. J. 44: 33–38.

Chhabra, R., and Singh,G. 1990. Technologies for afforestation of salt-affected soils of Indo-Gangetic plains. Proc. Natn. Symp. Technologies for afforestation of wastelands. Forest Research Institute, Dehradun, India.

Chhabra, R., Singh, M.V., and Abrol, I.P. 1982. Effect of method of zinc application on the yield of rice and wheat in a partially reclaimed sodic soil. Ann. Rep. Central Soil Salinity Research Institute, Karnal, India. p. 34.

Chhabra, R., Singh, N.T., and Hansen, A. 1989. Interaction between type of amendment and P availability. Ann. Rep. Central Soil Salinity Research Institute, Karnal, India. p. 38.

Chhabra, R., Singh, S.B., and Abrol, I.P. 1979. Effect of exchangeable sodium percentage on the growth, yield and chemical composition of sunflower (*Helianthus annuus* L.). Soil Sci. 127: 242–247.

Central Soil Salinity Research Institute. 1975. India salt-affected soils. Scale 1:6666 667. Karnal.

Crocker, C.D. 1973. General soil map of Kingdom of Cambodia. USAID/Ministry of Agriculture, Phnom Penh. 1963. In: Soil classification in tropical Asian Countries (in Japenese). National Institute of Resources, Japanese Science and Technology Agency, Tokyo, Japan.

Dahiya, S.S., and Singh, M. 1976. Effect of salinity, alkalinity and iron application on the availability of iron, manganese, phosphorus and sodium in pea (*Pisum sativum* L.) crop. Pl. Soil. 44: 697–702.

Dalton, F.N., Herkelrath, W.N., Rawlins, D.S., and Rhoades, J.D. 1984. Time-domain reflectrometry: Simultaneous measurements of soils water content and electric conductivity with a single probe. Science. 224: 989–990.

Dargan, K.S., Chandra, H., and Bhumbla, D.R. 1973. In saline-sodic soil—urea spraying for more paddy. Indian Fmg. 22(10): 9–10.

Dargan, K.S., Chillar, R.K., and Bhardwaj, K.K.R. 1975. In alkali soils—green manuring for more paddy. Indian Fmg. 25(3): 13–15.

Dargan, K.S., and Gaul, B.L. 1974. Optimum nitrogen levels for dwarf varieties of paddy in semi-reclaimed alkali soils. Fertil. News. 19(10): 24–25.

Dargan, K.S., Gaul, B.L., Abrol, I.P., and Bhumbla, D.R. 1976. Effect of gypsum, farmyard manure and zinc on the yield of barseem, rice and maize grown in a highly sodic soil. Indian J. Agric. Sci. 46: 535–541.

Dhillon, S.K., Sinha, M.K., and Randhawa, N.S. 1975. Chemical equilibria and Q/I relationship of zinc in selected alkali soils of Punjab. J. Indian Soc. Soil Sci. 23: 38– 46.

Dieleman, P.J. 1963. Reclamation of salt-affected soils in Iraq. Intern. Institute for Land Reclamation and Improvement, Wageningen, The Netherlands. Pub.II : 175.

Doering, E.J. 1963. A direct method for measuring the upward flow of water from the water table. Soil Sci. 96: 191–195.

Doering, E.J., and Sandoval, F.M. 1976. Hydrology of saline seeps in the Northern Great plains. Trans. ASEA. 19: 856–861.

Dregne, H.E. (ed.). 1977. Managing Saline Water for Irrigation. Proc. Intern. Conf. Managing Saline Water for Irrigation: Planning for Future. Intern. Center for Arid and Semi-arid Land Studies. Texas Tech. Univ. p. 618.

Dregne, H.E., and Mojallali, H. 1969. Salt fertilizer specific ion interaction in soils. New Mexico State Univ. Agric. Experiment Station. Bull. 541: 16.

Dubey, S.K., Mondal, R.C., and Gupta, S.K. 1988. Some aspects of water management in pitcher irrigation. Proc. Natn. Symp. Management of Irrigation System, Central Soil Salinity Research Institute, Karnal. pp. 25–27.

Dudal, R., and Purnell, M.F. 1986. Land resources: Salt-affected soils. Reclamation and Revegetation Res. 5: 1–9.

Duston, G.H., and Lonsford, J.V. 1955. Cannery waste disposal by irrigation. Sewage Industrial Waste. 21: 827.

Dutt, G.R., Terkeltoub, R.W., and Reuschkolb, R.S. 1972. Prediction of gypsum and leaching requirements for sodium affected soils. Soil Sci. 114: 93–102.

Eaton, F.M. 1935. Boron in soils and irrigation waters and its effect on plants with particular reference to the San Joaquan Valley of California. US Dept. Agric. Tech. Bull. 448: 131.

Eaton, F.M. 1950. Significance of carbonates in irrigation waters. Soil Sci. 69: 123–133.

El-Elgabaly, M.M. 1955. Specific effects of absorbed ions on plant growth. 1. Effect of different combinations of calcium, magnesium and sodium on barley seedlings. Soil Sic. 80: 235–248.

El-Elgabaly, M.M. 1959. Improvement of soils, irrigation and drainage in Egypt. The Supreme Council for Research (in Arabic).

El-Elgabaly, M.M. 1971. Reclamation and management of salt-affected soils. FAO. Irrigation and Drainage Paper. 17: 50–79.

El-Mahi, Y.E., and Mustafa, M.A. 1980. The effect of electrolyte concentration and sodium adsorption ratio on phosphate retention by soils. Soil Sci. 130: 321–325.

El-Swaify, S.A., Ahmed, S., and Swindale, L.D. 1970. Effects of adsorbed cations on physical properties of tropical red and tropical black earths. II Liquid limit, degree of dispersion and moisture retention. J. Soil Sci. 21 : 188–198.

Emerson, W.W., and Chi, C.L. 1977. Exchangeable calcium, magnesium and sodium and the dispersion of illites in water. Aust. J. Soil Res. 15: 255–262.

Fang, S., Tian, Y., and Xin, D. 1978. Comprehensive control of drought, waterlogging, salinization and saline groundwater. Selected Works of Symp. Reclamation of Salt-affected Soils in China. The Shandong Publishing House of Scientific Technology, China.

FAO. 1974. FAO-Unesco Soil Map of the World, Vol. I. Unesco, Paris.

FAO. 1977. FAO-Unesco Desertification Map of the World. Unesco, Paris.

Ferguson, W.S., and Herlin, R.A. 1963. Effect of soluble salts on plant response to and adsorption of phosphorus. Canad. J. Soil Sci. 43: 210–218.

Finck, A. 1977. Soil salinity and plant nutritional status. Proc. Intern. Symp. Managing Saline Water for Irrigation. Lubbock, Texas, pp. 199-210.

Fine, L.P., and Carson, P.L. 1954. Phosphorus needs of small grains on a moderately saline soil. Soil Sci. Soc. Am. Proc. 18: 60–63.

Fisk, W.W. 1964. Food processing waste disposal. Water and Sewage Works. 3: 417.

Fleisher, Z., and Hagin, J. 1981. Lowering ammonium volatilization losses from urea application by activation of nitrification process. Fert. Res. 2: 101–107.

Folkman, Y., and Wachs, A.M. 1973. Nitrogen removal through ammonia release from holding ponds. In Advances in Water Pollution Research, Ed. S.H. Jenkins Pergamon Press, New York.

Fournier, F. 1965. The salt-affected soils of the African continent. Proc. Symp. Sodic Soils, Budapest. Talej. es. Agrok. 14: 243–245.

Francois, L.E., and Bernstein, L. 1964. Salt tolerance of safflower. Agron. J. 56: 38–40.

Gill, H.S., and Abrol, I.P. 1985. Effect of posthole filling mixture composition on the growth and survival of tree species in sodic soils. Ann. Rep. Central Soil Salinity Research Institute, Karnal, India. pp. 26–28.

Grahm, P.H., and Parker, C.A. 1964. Diagnostic features in the characterization of the root nodule bacteria of legumes. Pl. Soil. 20: 383–396.

Gupta, I.C. 1979. Use of saline water in agriculture in arid and semi-arid zone of India. Oxford & IBH Pub. Co. New Delhi, p. 210.

Gupta, I.C. 1982. Soil salinity and lithium toxicity. Trans. Indian Soc. Desert. Technol. 7: 45–49.

Gupta, I.C., and Chandra, H. 1972. Effect of gypsum in reducing boron hazards of saline waters and irrigated soils. Ann. Arid Zone. 11: 228–230.

Gupta, R.K., Chhabra, R., and Abrol, I.P. 1982. Fluorine adsorption behaviour in alkali soils: Relative role of pH and sodicity. Soil Sci. 133: 364–368.

Gupta, S.K. 1982. Hydrological behaviour of alkali and normal soils: A comparison. J. Inst. Engrs. (India). 63: 18–20.

Hale, V.K., and Wallace, A. 1960. Bicarbonate and phosphorus effects on uptake and distribution in soybeans of iron chelated with ethylene diamine di-o-hydroxyphenyl acetate. Soil Sci. 89: 285–287.

Haque, Anwar, Ul 1989. Water Quality criteria based on an evaluation. Working Document 1. For Program Advisory Meeting 29—31 May, 1989. International Waterlogging and Salinity Research Institute, Lahore, Pakistan.

Hassan, N.A.K., Drew, J.V., Knudsen, D., and Olsen, R.A. 1970. Influence of soil salinity on production of dry matter and uptake and distribution of nutrients in barley and corn. I. Barley (*Hordeum vulgare* L.). Agron. J. 62: 43–45.

Hayward, H.E., and Spurr, W.B. 1944. Effects of iso-osmotic concentrations of inorganic and organic substances on the entry of water in corn roots. Bot. Gaz. 106: 131–139.

Hemminga, M.A., and Toorn, J.V.D. 1970. Flevobericht No. 73. Rijksdienst voor de Ijsslmeer polders. Zwolle. The Netherlands.

Hira, G.S., Bajwa, M.S., and Singh, N.T. 1981. Prediction of water requirement for gypsum dissolution in sodic soils. Soil Sci. 131(6): 353–358.

Hira, G.S., and Singh, N.T. 1980. Irrigation water requirement for dissolution of gypsum in sodic soils. Soil Sci. Soc. Am. J. 44: 930–933.

Hoffman, G.J. 1980. Irrigation management—salinity control. Proc. ASAE, 2nd Natl. Irrigation Symp., Irrigation Challenges of the 80's. Univ. Nebraska, Lincoln, Nebraska. pp. 166–174.

Hoffman, G.J., and van Genuchten, M. Th. 1983. Soil properties and efficient water use: Water management for salinity control. In Limitations to Efficient Water Use in Crop Production, Eds. H.M. Taylor, W. Jorden and T. Sinclair. Am. Soc. Agron. pp. 73–85.

Israelson, O.W., and Hansen, V.E. 1967. Irrigation Principles and Practices. John Wiley and Sons Inc., New York.

Janardhan, K.V., and Rao, C.V. 1970. Effect of soil salinity on the uptake of fertilizer phosphorus by rice. Oryza. 7: 93–94.

Jewitt, T.N. 1942. Loss of ammonia from ammonium sulphate applied to alkaline soils. Soil Sci. 54: 401–409.

Joshi, Y.S., Qadar, A., Bal, A.R., and Rana, R.S. 1980. Sodium/potassium index of wheat seedlings in relation to sodicity tolerance. Proc. Intern. Symp. on Management of Salt-affected Soils. Karnal, India. pp. 457–460.

Juwarkar, A.S. 1988. Reuse of waste waters for crop irrigation. Jal Vigyan Sameeksha. III: 82–96.

Kahane, J., and Mayber, A.P. 1968. Effect of substrate salinity on the availability for protein synthesis in pea roots. Pl. Physiol. 43: 1115–1119.

Kanwar, J.S. 1961. Clay minerals in saline and alkali soils of the Punjab. J. Indian Soc. Soil Sci. 9: 35–40.

Kanwar, J.S., Bhumbla, D.R., and Singh, N.T. 1965. Studies on the reclamation of saline and sodic soils in the Punjab, India. Indian J. Agric. Sci. 35: 43–51.

Kanwar, J.S., and Kanwar, B.S. 1969. Quality of irrigation waters. Trans. 9th Intern. Congr. Soil Sci. I: 391–403.

Kanwar, J.S., and Randhawa, N.S. 1974. Micronutrient Research in Soils and Plants. Indian Council of Agricultural Research, New Delhi. p. 185.

Kanwar, J.S., Sehgal, J.L., and Bhumbla, D.R. 1963. Relationship between some indices of saline sodic soils. J. Indian Soc. Soil Sci. 11: 39–44.

Kanwar, J.S., and Singh, S.S. 1961. Boron in normal and saline alkali soils of the irrigated areas of the Punjab. Soil Sci. 92: 207–211.

Katyal, J.C., Randhawa, N.S., and Sharma, B.D. 1980. All India Coordinated Scheme of Micronutrients in Soils and Plants. ICAR, New Delhi.

Katyal, J.C., and Sharma, B.D. 1980. A new technique of plant analysis to resolve iron chlorosis. Pl. Soil. 55: 105– 119.

Kaziev, M.V. 1956. The effect of adsorbed bases on the mobility of phosphates in Sierozems. Trudy S. Kh. Inst. 7: 49–52.

Kelley, W.P. 1951. Alkali Soils. Reinhold Pub. Corp., New York.

Khalil, M.A., Amer, F., and El-Elgabaly, M.M. 1967. A salinity fertility interaction study on corn and cotton. Soil Sci. Soc. Am. Proc. 31: 683–686.

Khosla, B.K., Dargan, K.S., Abrol, I.P., and Bhumbla, D.R. 1973. Effect of depth of mixing gypsum on soil properties and yield of barley, rice and wheat grown in a saline-sodic soil. Indian. J. Agric. Sci. 43: 1024–1031.

Khosla, B.K., Gupta, R.K., and Chawla, K.L. 1980. Evaluating the field hydraulic conductivity and soil salinisation under conditions of high water tables. Intern. Symp. Management of Salt-affected Soils. Karnal, India. pp. 256– 264.

Klein, G.B. 1986. Prcblem soils as potential areas for adverse salt-tolerant rice varieties in South and Southeast Asia. IRRI Research Papers Series No. 119. Nov. 1986. The International Rice Research Institute, Manila, Philippines. p 53.

Kolarkar, A.S., and Singh, N. 1970. Diagnosis of sodic conditions in saline alkali soils of Western Rajasthan. Ann. Arid Zone. 9: 31.

Kovda, V.A. 1965. Alkaline soda-saline soils. Proc. Symp. Sodic Soils, Budapest. Talaj. es. Agrok. 14: 15–48.

Kovda, V.A. 1973. (ed.). Irrigation, Drainage and Salinity—An International Source Book, FAO, Unesco, Rome. pp. 77–79.

Kruse, H. 1962. Some present day sewage treatment methods in use of small communities in Federal Republic of Germany. Bull. WHO. 26: 542.

Kumar Ashok, and Abrol, I.P. 1984. Studies on the reclaiming effect of Karnal grass and para grass grown in a highly sodic soil. Indian J. Agric. Sci. 54: 189–193.

Kumar, S., and Garg, O.P. 1980. Effect of saline-alkaline conditions on nodulation in pea (*Pisum sativum* L.). Indian J. Pl. Physiol. 23: 55–60.

Kumarswamy, S., and Mosi, A.D. 1969. Studies on fixation and availability of phosphorus in the alkaline soils of Tamilnadu. Madras Agric. J. 56: 351–359.

Lakshmi Kumari, M., Singh, C.S., and Suba Rao, N.S. 1974. Root hair infection and nodulation in lucerne (*Medicago salvia* L.) as influenced by salinity and alkalinity. Pl. Soil. 40: 261–268.

Langdale, G.W., and Thomas, J.R. 1971. Soil salinity effects on absorption of nitrogen, phosphorus and protein synthesis by coastal bermuda grass. Agron. J. 63: 708–711.

Luken, H. 1962. Saline soils under dryland agriculture in South Eastern Saskatchewan (Canada) and possibilities for their improvement. Part II. Evaluation of effects of various treatments on soil salinity and crop yields. Pl. Soil. 17: 26–48.

Lunin, J., and Gallatin, M.H. 1965. Salinity-fertility interactions in relation to the growth and composition of beans. I. Effect of N, P and K. Agron. J. 57: 339–342.

Maas, E.V., and Hoffman, G.H. 1977. Crop salt tolerance evaluation of existing data. Intern. Symp. Managing Saline Water for Irrigation, Lubbock, Texas, pp. 187– 198.

Maas, E.V., Ogata, G., and Garber, M.J. 1972. Influence of salinity on Fe, Mn and Zn uptake by plants. Agron. J. 64: 793–795.

Manchanda, H.R. 1976. Quality of groundwaters of Haryana. Bull., Haryana Agricultural University, Hisar, India.

Manchanda, H.R., and Sharma, S.K. 1983. Interactive effects of chloride and sulphate dominant soil salinities and phosphorus in relation to wheat. Tropical Pl. Sci. Res.

Margat, J. 1961. Les eaux salis en Maroc. Intern. Symp. Teheran. Unesco Publ. pp. 91–104.

Martin, J.P., and Bingham, F.T. 1954. Effect of various exchangeable cation ratios in soil on growth and chemical compositions of avocado seedlings. Soil Sci. 78: 349– 360.

Martin, J.P., Buehrer, T.F., and Caster, A.B. 1942. Threshold pH value for nitrification of ammonia in desert soils. Soil Sci. Soc. Am. Proc. 7: 223–228.

Massoud, F.I. 1974. Salinity and alkalinity as soil degradation hazards. FAO-Unesco Pub., Rome. 74: 10.

McClang, G., and Frankenberger, W.T. 1985. Soil nitrogen transformations as affected by salinity. Soil Sci. 139: 405–411.

McIntyre, D.S. 1979. Exchangeable sodium, soil plasticity and hydraulic conductivity of some Australian soils. Aust. J. Soil Res. 17: 115–120.

McNeal, B.L., Pearson, G.A., Hatcher, J.T., and Bower, C.A. 1966. Effect of rice culture on reclamation of sodic soils. Agron. J. 58: 238–240.

Mehrotra, C.L., and Das, S.K. 1973. Influence of exchangeable sodium on the chemical composition of important crops at different stages of growth. J. Indian Soc. Soil Sci. 21: 355–365.

Metcalf, L., and Eddy, H.P. 1972. Wastewater Engineering. McGraw-Hill Book Co. New York.

Migunova, E.S. 1976. The root systems of woody species on saline soils in the South Ukraine. Lesorren Devie. 6: 27–36.

Milap Chand, Bhumbla, D.R., Randhawa, N.S., and Sinha, M.K. 1980. The effect of gypsum and organic amendments on the availability of zinc to rice and its uptake by this crop grown in a saline sodic soil. Proc. Intern. Symp. On Management of Salt-affected Soils. Karnal, India. pp. 348– 355.

Ministry of Agriculture and Fisheries, Soil Science Division. 1970. Generalized soil map of West Malaysia 1970. Scale 1:760 320. Kuala Lumpur, Malaysia.

Mishra, S.G., and Pandey, G. 1976. Zinc in saline and alkali soils of Uttar Pradesh. J. Indian Soc. Soil Sci. 24: 336– 338.

Mondal, R.C. 1974. Farming with pitcher: A technique of water conservation. World Crops. 26: 94–97.

Muir, A. 1951. Notes on the soils of Syria. J. Soil Sci. 2: 163–182.

Mulder, E.C., and van. Veen, W.L. 1960. Effect of pH and organic compounds on nitrogen fixation by red clover. Pl. Soil. 13: 91–113.

Narayana, V.V.D. 1979. Rain water management for lowland rice cultivation in India. J. Irrigation and Drainage Division, ASCE. 105 (IRI): 87–98.

Nazir, A. 1965. A review of salinity-alkalinity status of irrigated soils in West Pakistan. Sodic Soils Symp. Baghdad, Talij. es. Agrok. 14:117–154.

Nightingale, G.T., and Farnham, R.B. 1936. Effect of nutrient concentration on anatomy, metabolism and abscission of sweet pea. Bot. Gaz. 97: 477– 917.

Nitant, H.C., and Bhumbla, D.R. 1974. Transformation and movement of nitrogen fertilizers in sodic soils. Ann. Rep. Central Soil Salinity Research Institute, Karnal, India. p. 24.

Nitant, H.C., and Dargan, K.S. 1974. Influence of nitrogenous fertilizers on yield and nitrogen uptake of wheat in a saline-sodic soil. J. Indian Soc. Soil Sci. 22: 121– 124.

Noggle, J.C. 1966. Ionic balance and growth of sixteen plant species. Soil Sci. Soc. Am. Proc. 30: 763–766.

Obrejanu, G., and Sandu, G.A. 1971. Amelioration of solonetz and solonet-zized soils in the Socialist Republic of Romania. In European Solonetz Soils and Their Reclamation. Edl. Szabolcs, Akademias Kiado, Budapest. pp. 99-130.

Oster, J.D., and Halvorson, A.D. 1978. Saline seep chemistry. In Dryland Saline Seep Control. Proc. 11th Intern. Cong. Soil Sci. Soc., Edmonton, Canada, pp. 27–29.

Oster, J.D., Hoffman, G.J., and Robinson, F.E. 1984. Management alterna-tives: Crop, water and soil. California Agriculture, October 1984: 29–32.

Pal, D.K., and Bhargava, G.P. 1979. Degradation of clay minerals in sodic soils. J. Indian Soc. Soil Sci. 27: 31–35.

Pal, D.K., and Mondal, R.C. 1980. Crop response to potassium in sodic soils in relation to potassium release behaviour in salt solution. J. Indian Soc. Soil Sci. 28: 347–354.

Paliwal, K.V. 1972. Irrigation with saline water. Monograph 2. Indian Agricultural Research Institute, New Delhi, India.

Pasricha, N.S., and Ponnamperuma, F.N. 1976. Influence of salt and alkali on ionic equilibria in submerged soils. Soil Sci. Soc. Am. J. 40: 374–376.

Pasricha, N.S., and Randhawa, N.S. 1971. Available molybdenum status of some recently reclaimed saline-sodic soils and its effect on concentration of molybdenum, copper, sulphur and nitrogen in barseem (*Trifolium alexandrinum*) grown on these soils. Proc. Intern. Symp. Soil Fertility Evaluation, Hisar, India. pp. 1017–1026.

Patel, P., and Wallace, A. 1975. Effect of mixed salinity and P fertility on yield and micro-element composition of tomato, sweet corn and sudan grass grown in sand culture. Indian J. Agron. 20: 29–34.

Pearson, G.A. 1960. Tolerance of crops to exchangeable sodium. US Dept. Agric. Inform. Bull. p. 216.

Pearson, G.A., and Bernstein, L. 1958. Influence of exchangeable sodium on yield and chemical composition of plants. II Wheat, barley, oats, rice, tall fescue and tall wheat grass. Soil Sci. 86: 254–261.

Pluym, H. van der. 1978. Extent, causes and control of dryland saline seepage in the northern great plains regions of North America. Proc. Dryland Saline Seep Control Meeting, Edmonton, Canada. pp. 48–58.

Pons, L.J. 1973. Outline of genesis, characteristics, classification and improvement of acid sulphate soils. In Acid Sulphate Soils. Ed.H. Doest. Proc. Intern. Symp. Wageningen, The Netherlands. ILRI Publ. 18(2): 3–23.

Polynov, B.B. 1930. Determination of critical depth of occurrence of the groundwater level salinising soils. Sector hydrotechnics and hydrotechnical construction No. 22, Leningrad.

Poonia, S.R., and Bhumbla, D.R. 1972. Effect of ESP on the availability of Ca from soil and added gypsum to maize (*Zea mays*) and dhaincha (*Sesbania aculeata*). Pl. Soil. 36: 671–679.

Pratt, P.F., and Thorne, D.W. 1948. Solubility and physiological availability of phosphate in sodium and calcium system. Soil Sci. Soc. Am. Proc. 13: 213–217.

Puttaswamygowda, B.S., and Pratt, P.F. 1973. Effect of straw, calcium chloride and submergence on a sodic soil. Soil Sci. Soc. Am. Proc. 37: 208–212.

Quidez, B., Bernardo, L. and Ponnamperuma, F.N. 1980. Coastal saline soils in the Philippines as potential rice lands. Paper presented at a Saturday Seminar. International Rice Research Institute, Los Banos, Laguna, Philippines.

Quirk, J.P., and Schofield, R.K. 1955. The effect of electrolyte concentration on the soil permeability. J. Soil Sci. 6: 163–178.

Rao, D.L.N., and Batra, Lalita. 1983. Ammonia volatilization from applied nitrogen in alkali soils. Pl. Soil. 70: 219–228.

Raychaudhuri, S.P. 1965. Classification and management of saline and alkali soils of India. Proc. Symp. Sodic Soils. Budapest. Talaj. es. Agrok. 14: 83–90.

Ressel, J.C., Kadry, L., and Hanna, A.B. 1965. Sodic soils in Iraq. Proc. Symp. Sodic Soils. Budapest. Talaj. es. Agrok. 14: 91–98.

Rhoades, J.D. 1972. Quality of water for irrigation. Soil Sci. 113: 277–284.

Rhoades, J.D. 1982. Reclamation and management of salt-affected soils after drainage. Proc. First Ann. Western Provincial Conf., Rationalization of Water and Soil Research and Management, Lethbridge, Alberta, Canada. pp. 123–197.

Rhoades, J.D., and Ingvalson, R.D. 1971. Determining salinity in field soils with soil resistance measurements. Soil Sci. Soc. Am. Proc. 35: 54–60.

Rhoades, J.D., and van Schilfgaarde, J. 1976. An electric conductivity probe for determining soil salinity. Soil Sci. Soc. Am. J. 40: 647–651.

Richards, L.A. (ed.) 1954. Diagnosis and Improvement of Saline Alkali Soils. USDA Handbook No. 60, Washington. p. 160.

Richards, L.A. 1966. A soil salinity sensor of improved design. Soil Sci. Soc. Am. Proc. 30: 333–337.

Rowell, D.L., and Shainberg, I. 1979. The influence of magnesium and of easily weathered minerals on hydraulic conductivity changes in a sodic soil. J. Soil Sci. 30: 719–726.

Ryden, J.C., and Syers, J.K. 1975. Rationalisation of ionic strength and cation effects on phosphate sorption by sodic soils. J. Soil Sci. 26: 395–406.

Sachin, R.S., and Menon, R.G. 1976. Use of sugar factory effluent for irrigation of sugarcane crop. Proc. Sem. Treatment and disposal of effluent from sugar and distillery industries. The Sugar Technology Association, India. Sugar Section. p. 31.

Sameni, A.M., Moftoun, M., Bassiri, A., and Sepaskhah, A.R. 1980. Growth and chemical composition of dry beans as affected by soil salinity and N-fertilization. Pl. Soil. 54: 217– 222.

Sandhu, S.S., and Abrol, I.P. 1981. Growth responses of *Eucalyptus tereticornis* and *Acacia nilotica* to selected cultural treatments in a highly sodic soil. Indian J. Agric. Sci. 51: 437–443.

Schlesener, B.A. 1958. Factors affecting evaporation from soils in contact with a water table. Ph.D. thesis, Colorado State University, Fort Collins, Colorado.

Schoonover, W.R. 1952. Examination of soils for alkali. University of California, Extension Service, Berkeley, California.

Sehgal, J.L., and Sys, C. 1980. Criteria for mapping salt-affected soils of the Mesopotamian and Indo–Gangetic plains. Proc. Symp. Salt-affected Soils, Karnal, India. pp. 56–66.

Shahi, H.N., Khind, C.S., and Gill, P.S. 1976. Iron chlorosis in rice (*Oryza sativa* L). Pl. Soil. 44: 231–232.

Shalhevet, J., and Bernstein, L. 1968. Effect of vertical heterogeneous soil salinity on plant growth and water uptake. Soil Sci. 106: 85–93.

Sharma, D.L., Moghe, V.B., and Mathur, C.M. 1968. Salinity and alkalinity problem and fertility status of soils of Pali district (Rajasthan). J. Indian Soc. Soil Sci. 16: 263– 269.

Sharma, D.P., Singh, K.N., and Rao, K.V.G.K. 1989. Reuse of drainage effluent. Ann. Report. Central Soil Salinity Research Institute, Karnal, India.

Sharma, O.P., and Gupta, R.K. 1986. Comparative performance of gypsum and pyrites in sodic vertisols. Indian J. Agric. Sci. 56: 423–429.

Sharma, S.K. 1980. Effect of different rates and methods of nitrogen application on yield of egg plant under saline conditions of Yemen Arab Republic. Indian J. Agron. 25: 557–558.

Shukla, U.S., Mittal, S.B., and Gupta, R.K. 1980. Zinc adsorption in some soils as affected by exchangeable cations. Soil Sci. 129: 366–370.

Sidhu, P.S., and Gilkes, R.J. 1977. Mineralogy of soils developed on alluvium in the Indo-Gangetic plains (India). Soil Sci. Soc. Am. J. 41 : 1194–1201.

Singh, A., Chhabra, R., and Abrol, I.P. 1979a. Effect of fluorine and phosphorus on the yield and chemical composition of rice (*Oryza sativa* L.) grown in soils of two sodicities. Soil Sci. 127: 86–93.

Singh, A., Chhabra, R., and Abrol, I.P. 1979b. Effect of fluorine and phosphorus applied to a sodic soil on their availability and on yield and chemical composition of wheat. Soil Sci. 128: 90–97.

Singh, B., Bhumbla, D.R., and Randhawa, N.S. 1975. Forms of inorganic phosphate in sodic soils. J. Indian Soc. Soil Sci. 23: 353–356.

Singh, C.S., Lakshmi Kumari, M., Biswas, A., and Subha Rao, N.S. 1973. Effect of carbonate and bicarbonate of sodium on growth of Rhizobia and nodulation in lucerne (*Medicago sativa* L.). Indian J. Microbiol. 13: 125–128.

Singh, D., and Nijhawan, S.D. 1943. Availability of phosphates in alkaline and calcareous soils. Indian J. Agric. Sci. 13: 134–141.

Singh, G., and Abrol, I.P. 1986. Agronomic investigations on production of *P. juliflora* under highly alkali soil conditions. Ann. Rep. Central Soil Salinity Research Institute, Karnal, India. pp. 98–107.

Singh, G., Sharma, D.P., and Mehta, K.K. 1983. Effect of sources, levels and methods of nitrogen application on yield attributes, yield and N uptake of rice in a highly sodic soil. 46th Ann. Conv. Indian Soc. Soil Sci., Hisar, India.

Singh, J., and Randhawa, N.S. 1977. Boron fractionation and mineral composition of saline alkali soils. J. Indian Soc. Soil Sci. 25: 433–435.

Singh, M. 1970. Distribution of manganese in tropical arid brown sierozem and desert soils of earstwhile Punjab (India). Agrochemica 14: 565–576.

Singh, M.V., Chhabra, R., and Abrol, I.P. 1979. Effect of levels of zinc on the yield of crops in sodic soils. Ann. Rep. Central Soil Salinity Research Institute, Karnai, India. pp. 33–34.

Singh, M.V., Chhabra, R., and Abrol, I.P. 1984. Factors affecting DTPA extractable Zn in sodic soils. Soil Sci. 136: 359–366.

Singh, M.V., Chhabra, R., and Abrol, I.P. 1987. Interaction between application of gypsum and zinc sulphate on the yield and chemical composition of rice grown on an alkali soil. J. Agric. Sci. Camb. 108: 275–279.

Singh, N.T. 1992. Salt-affected soils in india. In: Land and Soil. T.N. Khoshoo and B.L. Deekshatulu (Ed). Har-Anand Publications, New Delhi, India. pp. 65–100.

Singh, N.T. 1974. Physico-chemical changes in sodic soils incubated at saturation. Pl. Soil. 40: 303–311.

Singh, N.T., Bhumbla, D.R., and Kanwar, J.S. 1969. Effect of gypsum alone and in combination with plant nutrients on crop yield and amelioration of a saline sodic soil. Indian J. Agric. Sci. 39: 1–9.

Singh, N.T., Hira, G.S., and Bajwa, M.S. 1981. Use of amendments in reclamation of alkali soils in India. Agrok. es. Talaj. 30: 158–177.

Singh, S., and Singh, S.B. 1975. Effect of application of iron and manganese on their uptake and yield of rice. J. Indian Soc. Soil Sci. 23: 489–493.

Singh, S.B., Chhabra, R., and Abrol, I.P. 1979. Effect of exchangeable sodium on the yield and chemical composition of raya (*Brassica juncea* L.). Agron. J. 71: 767–770.

Singh, S.B., Chhabra, R., and Abrol, I.P. 1980. Effect of soil sodicity on the yield and chemical composition of cowpea (*Vigna unguiculata*) grown for fodder. Indian J. Agric. Sci. 50: 852–856.

Singh, S.B., Chhabra, R., and Abrol, I.P. 1981. Effect of exchangeable sodium on the yield, composition and oil content of safflower (*Carthamum tinctorius* L.) and linseed (*Linum usitatissimum* L.). Indian J. Agric. Sci. 52: 881–891.

Soil Research Institute. 1976. Generalized soil map of Indonesia. Scale. 1:2500, 000. 4th Edition. Bogor, Indonesia.

Stoneman, T.C. 1978. The nature and extent of salinity problems in Western Australia. In Dryland Saline Seep Control. 11th Congr. Intern. Soil Sci. Soc., Edmonton, Canada. Proc. Subcommission on Salt-affected Soils. pp. 30–35.

Subha Rao, N.S., Lakshmi Kumari, M., Singh, C.S., and Magu, S.P. 1972. Nodulation of lucerne (*Medicago sativa* L.), under the influence of sodium chloride. Indian J. Agric. Sci. 42: 384–386.

Swarup, A. 1980. Effect of submergence and farm yard manure application on the yield and nutrition of rice and sodic soil reclamation. J. Indian Soc. Soil Sci. 28: 532–534.

Szabolcs, I. 1965. Investigations with radioactive traces of sodium carbonate in soils. Proc. Symp. Isotope Radiation. Anakra. pp. 37–53.

Szabolcs, I. 1974. Salt-affected soils in Europe. Martinus Nijhoff, the Hague, The Netherlands. p. 63.

Szabolcs, I. 1979. Review of research on salt-affected soils. Natural Resources Research xv, Unesco, Paris.

Takkar, P.N., and Randhawa, N.S. 1978. Micronutrients in Indian Agriculture. Fert. News. 23(8): 3–26.

Takkar, P.N., and Sidhu, B.S. 1979. Kinetics of zinc transformation in submerged alkali soils in the rice growing tracts of Punjab. J. Agric. Sci. Camb. 93: 441–447.

Talsma, T. 1963. The control of saline groundwater. Meded 63, Landbouw hogeschool, Wageningen, The Netherlands. pp. 1–68.

Taylor, A.W., and Gurney, E.L. 1965. The effect of salts on the dissolution of monocalcium phosphate. Soil Sci. Soc. Am. Proc. 29: 151–153.

Thomas, J.R. 1980. Osmotic and specific salt effects on growth of cotton. Agron. J. 72: 407–412.

Thomas, J.R., and Langdale, G.W. 1980. Ionic balance in coastal bermuda grass as influenced by nitrogen fertilization and soil salinity. Agron. J. 72: 449–452.

Tomar, O.S., and Gupta, R.K. 1986. Performance of tree species in saline soils. Ann. Rep. Central Soil Salinity Research Institute, Karnal, India. pp. 119–123.

Torres, B.C., and Bingham, F.T. 1973. Salt tolerance of Mexican wheat. 1. Effect of NO_3 and NaCl on mineral nutrition, growth and grain production of four wheats. Soil Sci. Soc. Am. Proc. 37: 711–715.

Tyagi, N.K., Singh, O.P., Dhruvanaryana, V.V., and Gupta, R.K. 1980. Rational approach to groundwater exploitation by shallow tubewells in Karnal, India. Irrig. Power J. 37: 241–147.

Van Alphen, J.G. 1975. Salt-affected soils in Peru. International Institute for Land Reclamation and Improvement, Wageningen. Ann. Rpt. pp. 7–13.

Van der Molen, W.H. 1976. Natural factors. In Prognosis of Salinity and Alkalinity. Soil Bull. 31. FAO, Rome.

Van Hoai, Nyugen 1973. General soil map of Vietnam. Scale 1: 2000,000-1962. In Soil Classification in Tropical Asian Countries (in Jananese). National Institute of Resources, Japanese Science and Technology Agency, Tokyo, Japan.

Varallyay, G. 1977. Moisture status and flow phenomena of salts in salt-affected soils. Proc. Indo-Hungarian Sem. Management of Salt-affected Soils. Karnal, India. pp. 85–102.

Wadleigh, C.H., and Ayers, A.D. 1945. Growth and biochemical composition of bean plants as conditioned by soil moisture tension and salt concentration. Pl. Physiol. 20: 106–132.

Westerman, R.L., and Tucker, T.C. 1974. Effect of salts and salt plus nitrogen labelled ammonium chloride on mineralization of soil nitrogen, nitrification and immobilization. Soil Sci. Soc. Am. Proc. 38: 602–605.

Woodruff, J.R., Lingon, J.T., and Smith, B.R. 1984. Water table depth interaction with nitrogen rates in sub-irrigated corn. Agron. J. 76: 280–283.

Yadav, J.S.P., and Agarwal, R.R. 1959. Dynamics of soil changes in the reclamation of saline alkali soils of the Indo-Gangetic alluvium. J. Indian Soc. Soil Sci. 7: 213–222.

Yadav, J.S.P., and Girdhar, I.K. 1981. Effect of varying Mg/Ca ratio and electrolyte concentration in the irrigation water on the soil properties and growth of wheat. Pl. Soil. 56: 413–427.

Yadav, J.S.P., and Singh, K. 1970. Tolerance of certain forest species to varying degree of salinity and alkali. Indian Forester. 96 (8): 587–599.

Yadav, N.K., and Vyas, S.R. 1971. Response of root nodule Rhizobia to saline alkaline and acid conditions. Indian J. Agric. Sci. 41: 875–881.

Zavaleta, A.G. 1965. The nature of saline and alkali soils of the Peruvian Coastal Zone. Proc. Symp. Sodic Soils. Budapest. Talaj. es. Agrok. 14: 415–424.

Zhukovskaya, N.V. 1973. Uptake and accumulation of phosphate by plants in salinised soils. Soils and Fertil. 36: 241.

Authors Index

Subject Index

T - #0031 - 101024 - C20 - 229/152/17 [19] - CB - 9789054107279 - Gloss Lamination